Digital Worldbuilding and Ecological Readiness

ENVIRONMENTAL COMMUNICATION AND NATURE: CONFLICT AND ECOCULTURE IN THE ANTHROPOCENE

Series Editor: C. Vail Fletcher, University of Portland

This interdisciplinary book series seeks original proposals that examine environmental communication scholarship. In the Anthropocene era, the period during which human activity has become the dominant influence on climate and the environment, the need for highlighting and recentering nature in our worldviews and policies is urgent, as collapsing ecosystems across the globe struggle to survive. Topics might include climate change, land use conflict, water rights, natural disasters, nonhuman animals, the culture of nature, ecotourism, wildlife management, human/nature relationships, food studies, sustainability, eco-pedagogy, mediated nature, eco-terrorism, environmental education, ecofeminism, international development, and environmental conflict. Ultimately, scholarship that addresses the general overarching question "how do individuals and societies make sense of and act against/within/out of nature?" is welcomed. This series is open to contributions from authors in environmental communication, environmental studies, media studies, rhetoric, political science, critical geography, critical/cultural studies, and other related fields. We also seek diverse and creative epistemological and methodological framings that might include ethnography, content analysis, narrative and/or rhetorical analysis, participant observation, and community-based participatory research, among others. Successful proposals will be accessible to a multidisciplinary audience.

Recent Titles in the Series

Digital Worldbuilding and Ecological Readiness, by Cynthia Porter Rosenfeld
Social Media and Oil in Southern California: Greenwashing Los Angeles, by Jason L. Jarvis
Hyperlocal Organizing: Collaborating for Recovery Over Time, by Jack L. Harris

Digital Worldbuilding and Ecological Readiness

Cynthia Porter Rosenfeld

LEXINGTON BOOKS
Lanham • Boulder • New York • London

Published by Lexington Books
An imprint of The Rowman & Littlefield Publishing Group, Inc.
4501 Forbes Boulevard, Suite 200, Lanham, Maryland 20706
www.rowman.com

86-90 Paul Street, London EC2A 4NE

British Library Cataloguing in Publication Information Available

Library of Congress Cataloging-in-Publication Data

Names: Rosenfeld, Cynthia Porter, 1986– author.
Title: Digital worldbuilding and ecological readiness / Cynthia Porter Rosenfeld.
Description: Lanham : Rowman & Littlefield, [2023] | Series: Environmental communication and nature : conflict and ecoculture in the anthropocene | Includes bibliographical references and index.
Identifiers: LCCN 2023036547 (print) | LCCN 2023036548 (ebook) | ISBN 9781666915464 (cloth) | ISBN 9781666915471 (ebook)
Subjects: LCSH: Communication in ecology. | Ecology—Computer network resources. | Communication in community development. | Community development—Computer network resources.
Classification: LCC QH541.18 .R67 2023 (print) | LCC QH541.18 (ebook) | DDC 577.01/4—dc23/eng/20230926
LC record available at https://lccn.loc.gov/2023036547
LC ebook record available at https://lccn.loc.gov/2023036548

∞™ The paper used in this publication meets the minimum requirements of American National Standard for Information Sciences—Permanence of Paper for Printed Library Materials, ANSI/NISO Z39.48-1992.

To
Emma, The Domonique, and The Australorp;
All Earthlings—past, present, and future;
and Lawrence Rosenfeld—always already.

Contents

List of Tables

Acknowledgments

The problem with entanglement is the knowledge that my acknowledgments—how I gesture to the many voices and perspectives that animate the pages ahead—cannot be anything but incomplete. They will highlight some important contributions of which I am aware and obscure other contributions that helped this project come into being. Thus, I both celebrate and mourn this section: I rejoice in the pleasure of recognizing the community behind this project even as I regret its certain inadequacy.

First, my deepest gratitude to the communities I think with in this book: people who share their plant-based diets on social media, snake identification groups, the Pink Chicken Project, and the Ngalia community and collaborators of Reclaim the Void. I want to express a particular appreciation to everyone who took time to talk or otherwise share their wisdom with me. I have withheld identities and assigned pseudonyms to all of the individuals who participated in interviews with me, with the exception of my chickens.

Thank you to Dr. Vail Fletcher, Nicolette Amstutz, Deja Ryland, the reviewer of this manuscript, and everyone at Lexington Books who helped care this project into being.

It is my pleasure to celebrate the contributions of Drs. Victoria Gallagher, Elizabeth Craig, Stephen Wiley, John Morillo, and Matthew Calarco to this project. Thank you for being not just the committee of people but the community of people who gave this project life. As I like to tell people, when I started my Masters of Science in Communication in 2017, I imagined that I would do some research on social movements and later—much later—in my career, I would get to my commitments to the more-than-human world. To my surprise and delight, I was met by a community of people who—like the social work credo—met me where I was, and then helped me go where I wanted while showing me myriad and wondrous possibilities for ways to get there. Yours were welcome, guiding voices to have in my head while writing, and each of you has been part of my own flourishment in recent years. I hope this project rewards all the care you have invested.

There is an honorary member of this committee to whom I am also indebted, Dr. William Kinsella. In this book, I talk a great deal about transformation and how worlds get built—that's because I know a great deal about transformation and how worlds get built. My world of evidence-based practice based on randomized controlled clinical trials (bonus points if the trial was double blind) met your world of critical and interpretive inquiry and became *otherwise*. Although I take responsibility for the contents of this book, I could not have crafted this project without you.

To my many teachers, mentors, collaborators, and colleagues at North Carolina State University over the past years, all of you are inscribed in these pages in some way. This has been the most intellectually challenging and rewarding time of my life, and it would not have been possible without this community.

To my family, friends, and colleagues who offered bountiful social support throughout this journey, thank you. Thank you, Monica Allgauer, for being a companion on this trek since I first submitted an essay to a writing competition in high school, and thank you to COHORT-19 and the MS of Communication Class of 2019 for seeing me to this milestone. Thank you to my mentors and colleagues at *The Roanoke Times*, the UNC School of Social Work, the Humane Society of the United States, North Carolina Museum of Natural Sciences, and the network of North Carolina artists and artisans with whom we worked for expanding my horizons. Thank you, especially, to two departed mentors and friends, Teresa Ilinitch and Matthew Howard—their loss is significant because their impact was tremendous. Thank you to the Richmans, Cranes, Callanans, and Rennies for your steadfast nurturance. Thank you to my brother, sister-in-law, nephew, and niece for all the ways you made this book possible over the years, and thank you to the Jeffersons, Cussens, Rosenfelds, Wassermans, and Roimishers for all the ways you have made me part of your families. Thank you to Olivia, Malcolm, Asher, Gideon, Corbin, and Emily for being people who always help me to remember the joys of humanity.

Thank you, Stuart Rennie, for introducing me to philosophy and for being the consistently steady being you are. In so many ways, you helped me bring this book into being.

Thank you to oaks, maples, sweetgums, poplars, dogwoods, sourwoods, and pines; the devil's urns, Jackson's slender Caesars, chanterelles, false turkey tails, witch's butter, false deathcaps, and false Caesars; the hawks, snakes, squirrels, moles, songbirds, crows, slugs, lizards, deer, opossums, raccoons, and box turtles; and to seven rowdy caprines, one patient feline, twenty-two highly autonomous chickens, two gentle cavies, and three devoted canines: I may have written *a* book without living with all of you, but

I would not have written *this* book. I am so grateful for the everyday *becoming* with these beings.

Finally, Lawrence Rosenfeld: in this endeavor, and all others, you are my partner. My partner, my best friend, my confidant, my companion—all of these words speak truth, but they speak this truth insufficiently. It is doubtful I would have written this book or any other without you. Thank you for the world we have built together.

Chapter 1

Introduction

Cladoselache is a genus of extinct sharks. Traces of *Cladoselache* have been found in rock formations from the Upper Devonian period, between 359 and385 million years ago, in North America and Europe. Though *Cladoselache* is widely agreed to be an early ancient shark, it was certainly unique among Elasmobranchii fish (a subclass of cartilaginous fish that includes ancient as well as modern sharks, skates, rays, and sawfish). In some respects, the bodies of *Cladoselache* resembled modern members of the Lamnidae family, which includes the well-known white sharks and makos, but in other ways, *Cladoselache* is not only distinct from modern sharks but even other ancient sharks: *Cladoselache* lacked the tooth-like scales that cover the bodies of most sharks. This absence made *Cladoselache* vulnerable to all manner of injury, particularly from the large, bony predatory fish known as *Dunkleosteus*. Yet this absence was also part of an overall composition of body that allowed *Cladoselache* to flourish as high-speed predators themselves (*Britannica* 2013).

This book is not about sharks, ancient or modern, but it does concern the geologic record and is absolutely about the entanglement of vulnerability and flourishment. The geologic record of concern here is not millions of years old but is actively under construction and has been given the moniker *Anthropocene*. The Anthropocene refers to the remarkable, systemic-altering impact of human activity on the more-than-human world that will show up in the geologic strata through the increased presence of radioactive elements, plastic pollution, and amassed chicken bones. New geologic epochs occur without human intervention, of course, but they typically take place over millions of years. Although the timing of the onset of the Anthropocene is debated—ranging from the agricultural and industrial revolutions to the Second World War—what is not debated is that the massive anthropogenic transformations of the more-than-human world have occurred very rapidly (Lewis and Maslin 2018). Species are unable to adapt quickly enough to the rapidly changing climate and mass habitat destruction characteristic of the

Anthropocene, and humans are witnessing and occasioning what is called the *sixth mass extinction* (Kolbert 2014). The name *Anthropocene*—like all language—affords and constrains human understanding about the apparatuses of ecological destruction. The rhetorical force of the term comes from its readily identifiable locus: the *anthropos*, or humans. The term encourages critical reflection on human activity and anthropocentric worldviews. However, the term also erases distinctions between ways of being human. The significance of this erasure is difficult to overstate. When understanding the hegemonic performance of industrial human ways-of-life as *the* way of being human, people can come to the conclusion that the most ethical response to the Anthropocene is voluntary human extinction (discussed more in chapter 2). However, there have always been multiple ways of being human, and some of those ways consider ecology as essential to human wellbeing as economy and health.

Further, the term can give the false impression that this is the age of human flourishment—an era named to mark the capacity of humans to reshape their environment! In the Anthropocene, humans are not faring so well either. Indeed, nonhuman animals are not the only climate refugees, and the need for people to relocate due to drought and famine exacerbate already precarious international, political environments (Berchin et al. 2017). The changing climate also creates crises that are intranational, as wildfires (Flavelle and Popovich 2022) and coastal and inland flooding (Flavelle et al. 2020) raise questions about what parts of a country will be habitable for how long. The landing page of the *New York Times* website now features a daily pandemic tracker, alerting visitors as to the current infection, hospitalization, and death rate for the coronavirus. As I am writing this, Russia has invaded Ukraine, setting off a cascade of events—including supply chain shortages—that illuminate how globally entangled all aspects of modern life are (Strubenhoff 2022). In the United States, there are hearings underway to publicly document the events leading to the tragic consequences of the January 6, 2021, insurrection at the US Capitol, and as of June 2, 2022, there were more than 250 year-to-date mass shootings in the United States (Ledur and Rabinowitz 2022). Meanwhile, people in their everyday lives try to stay connected with loved ones and perhaps find some relief through technologies that contribute to massive energy consumption and waste-world-making (Hird 2016)—through the rapid use and disposal of phones, computers, tablets, and gaming systems—that are part of the mechanisms of the Anthropocene (Perkins et al. 2014).

Even within the digital spaces those technologies afford—ignoring for a moment their material consequences and the sociopolitical physical conditions in which these technologies and digital spaces exist—things are not great. Amid the current bear market of 2022-to-date and bank failures around the globe, there are also concerns about a cryptocurrency economic bubble and burst (Flitter 2022), not unlike the dot-com boom and bust that triggered the early 2000s recession, and this virtual currency is increasingly used in human and drug trafficking (Knutson 2022). Tech giants have ushered in a model of capitalism where the product is users' information and that product is sought not only by commercial advertisers but also by political actors (Zuboff 2019) and have both created and warned of the threat Artificial Intelligence poses to human survival (Roose 2023). Social media is a means to share pictures and updates with friends and family, but it is also a medium for doxing, harassing, threatening, trolling, and otherwise bullying people. People share horrific videos for the spectacle (Chouliaraki and Kissas 2018) and hurtful memes for the lolz (Woods and Hahner 2019). Even when these most ugly of interactions are not occurring, the curated identities of social media contribute to feelings of FOMO (fear of missing out), isolation, and physical, social, and occupational inadequacy (Roberts and David 2020).

All of this is true, and all of this makes the Anthropocene a challenging time to live in, let alone flourish. Yet, this book takes flourishment not only as a possibility but as central to questions of ethics, of what it means to live well with others. The capacity to recognize and analyze crises is essential but so is the capacity to recognize and analyze ecological wellbeing. Since 2015, several introductions to Western philosophy and psychology of wellbeing have been published, as have multiple edited volumes on wellbeing from Indigenous perspectives. The study of wellbeing has become more common, but it is still something of an outlier. Ecolinguistic scholars calling for studies of ecological wellbeing have called for studies of *positive discourse analysis* (Stibbe 2018), evincing an understanding that *discourse analysis*—without the modifier—is unlikely to engage affirmative content or produce an affirmative reading. In medical sciences, many people will be familiar with the study of *pathogenesis*, the study of disease etiology and progression. Fewer are likely to be familiar with the study of *salutogenesis*, the study of factors that promote and maintain wellbeing. Indeed, Microsoft Word had no trouble recognizing *pathogenesis* but placed a red, squiggly line under *salutogenesis* and recommended "autogenesis" in its place.

The flourishment discussed in this book recognizes desire fulfillment and pleasure as part of wellbeing but offers a more expansive vision than that. Flourishment, here, attends to both subjective and objective aspects of wellbeing (discussed more in chapter 2) and is informed by the work of feminist ethics of care theorists who seek to "maintain, continue, and repair our 'world' so that we can live in it as well as possible. That includes our bodies, our selves, and our environment, all of which we seek to interweave in a complex, life-sustaining web" (Tronto 1993, 103).

Flourishment involves caring for and caring about ourselves, each other, and the more-than-human world; it is "a mode of thriving—knowing, doing, and making—attuned to what an environment affords" (Rickert 2013, 15), and it is associated with joy. Joy, in Spinoza's *Ethics*, is expansive, compounds relations, and increases the power of living. Restoring joy is an ethical endeavor.

As discussed throughout this book, flourishment is a communal virtue that is partial, open, response-able/responsible, and contestable, but that nonetheless compels people to act. As a communal virtue of action, my conceptualization of flourishment also draws on Aristotle's conception of *eudaimonia*, or the virtue of flourishment. Lisa Tessman (2005) describes the spirit of *eudaimonism* as an affirmation and embrace of life, even in the dire circumstances of the Anthropocene. Eudaimonism is not a cruel optimism or toxic positivity, though. The eudaimonistic affirmation of life recognizes the entanglement of vulnerability and flourishment (168).

Although this book recognizes the interdependency of vulnerability and flourishment as inevitable and necessary, my method of travel for this journey is to take a largely affirmative account of what is being done *well* in human attempts to live well with others. Deleuze (1988) writes that according to Spinoza, "It is a disgrace to seek the internal essence of man in his bad extrinsic encounters. Everything that involves sadness serves tyranny and oppression" (72). Although the community-care paradigm developed in chapter 2 argues against any internal essence of humanity, I do agree that there is something *good*—serves to increase people's agency—in work that explores instances of living well, and it is *bad*—serves to decrease people's agency—if the only thing on offer is a view of humanity that leads to misanthropy.

In Spinoza's understanding of ethics, the paradigmatic example of bad relations is a poison, which works to decompose one's body. So here it is important to know what "poisons" are out there—from environmental racism to the marketization of all aspects of existence—and how they work. Deconstructionist and posthumanist scholarship, which illuminates the pervasiveness of a narrow form of anthropocentrism that oppresses humans and more-than-human beings alike, provides one avenue for recognizing these poisons and how they function. It is also important to know what tonics and remedies are out there and how they work, which is a project taken up well by decolonial (e.g., Haas 2015) and Black media theorists (e.g., Gallon 2016). With this book, I hope to contribute to the available "equipment for living" (Burke 2012) to further strengthen our stock of remedies. Together, the landscape of scholarship of poisons and tonics can aid our understanding of what rhetorics, relations, politics, economic arrangements, and social configurations best undermine and best contribute to (more-than-)human flourishment and how to build more just and caring worlds.

Two final notes before I map the journey ahead. First, I have constrained my search for tonics to digital discourses and projects with a significant—and in one case, entirely—digital presence. I elaborate on this decision in chapter 3, but for now, I want to acknowledge the impetus for this choice. As Johanna Hartelius (2021) demonstrated, digital worlds are also more-than-human worlds and are spaces that provoke questions of how to live well in and with them. The self-imposed constraint of seeking digital rhetorics that build more flourishing more-than-human worlds is a move to resist some of the techno-determistic discourse that can creep into environmental studies and prevent environmental studies from engaging with *all* of the more-than-human world. At the same time, the physical more-than-human world and its inhabitants are often overlooked in digital studies. New technologies bring about new hopes for radical change in human sociality—like the burgeoning realm of metaverses (Larsson 2022)—but what of the more-than-human world? How will the proliferation of metaverses act on the more-than-human world, and how will the more-than-human world show up in these metaverses? The purpose of this book is to examine how some digital spaces act on and with human diets, how lived relations with nonhuman animals can be transformed, and how conceptualizations of land are (re)made in order to offer some tools and tonics for navigating metaverses present and future.

Second, I want to offer a brief introduction to the language used in this book. Trying to write about the more-than-human world in English from a perspective of embeddedness and entanglement is often inelegant. The English language is much more conducive to writing from an understanding of nature/culture separation than from a recognition of natureculture's (Haraway 1998) inseparability. Throughout this book, I will use neologisms from new materialist theorists, like Donna Haraway and Karen Barad, and even contribute a couple of neologisms myself. I also try to preserve the language used in the digital rhetorics I analyze as much as possible, so—for example—in chapter 4, there will be several comments that are written in different languages. I endeavor to define or translate each new word on its initial use, but I want the reader to know that some backtracking may be necessary at times if one were to turn to a random page and begin reading.

Also, I will use the terms more-than-human and (more-than-)human a lot in this book. The *more-than-human world* is a term coined by David Abram (1996) to describe what has typically been referred to as "nature" or the "environment" but without invoking the separation inherent in these terms. The *more-than-human world* both includes and exceeds humans. The reason that I sometimes bracket (more-than-) is that there are times when I want to draw attention equally to how things show up in the material world exceeding humans *and* in human sociopolitical configurations. Even when I make use of the brackets, the always-already embeddedness of human affairs in the more-than-human world is implied.

Overall, the book is structured in a way that each chapter builds on the previous one *while* also affording a productive reading in isolation (with the caveat that some word choices may be a bit of a challenge to decode). For example, chapter 5 on snake identification communities could be read alone by someone interested in how the chapter takes up invitational rhetoric in digital spaces, but the chapter also builds on the preceding chapters and becomes part of the source material for the chapters following. Overall, each chapter of the book contributes to theorizing and illuminating how digital rhetorics can afford a sense of ecological readiness: a sense of being able to participate in crafting more-than-human worlds of contingent wellbeing, or vulnerable flourishment. Vulnerable flourishment is a touchstone that will be returned to throughout this book, as a reminder that flourishment and wellbeing are never absolute.

Chapter 2 lays out the ontological, ethical, and epistemological—or ethico-onto-epistemologies (Barad 2007)—foundations of what I call the community-care paradigm. The community-care paradigm, in which humans are understood as embedded in a more-than-human community that makes claims on us, is conceptualized through Aldo Leopold's (1987) land ethic and Phaedra Pezzullo's and J. Robert Cox's (2021) articulation of environmental communication as a crisis-care discipline; the ethico-onto-epistemologies of the community-care paradigm are also found to be in harmony with and enhanced by Indigenous cosmovisions and epistemologies (Chilisa 2019; Grim 2001). The community-care paradigm is theorized as a way of understanding and living in the more-than-human world that is based on a relational ontology, situated knowledges, and ethics of care that takes individuals-in-communities as its basic unit of consideration.

In chapter 3, I develop the idea of *worldbuilding rhetorics* by examining rhetorical concepts beginning in the late twentieth century that illuminate how rhetoric participates in bringing worlds into being. I introduce and use these concepts in an explication of a critical making project called Talking Trash. Talking Trash is a paper-woven trash receptacle with an associated Twitter account that seeks to make public the (seemingly) private act of waste disposal. Talking Trash illuminates the online-offline reverberations of worldbuilding and moves the chapter's conversation to the ways that digital spaces facilitate and foreclose possibilities for being-in-the-world.

The next three chapters all offer rhetorical worldbuilding concepts that I articulate through a grounded-theory development based on rhetorical field methods (Endres et al. 2016; McKinnon et al. 2016; Senda-Cook et al. 2018), which combine the techniques of digital ethnography (Pink et al. 2015; Pink

2016) with rhetorical criticism. Chapter 4 explores the worldbuilding capacities of *disclosure* as theorized jointly from interpersonal communication (i.e., self-disclosure as personal acts of revealing aspects of oneself) and rhetorical theory (i.e., the supra-intentional ways that aspects of the world are revealed). This accounting of disclosure helps clarify how vernacular plant-based visual rhetorics (i.e., digital food photography cataloged with plant-based hashtags) on Instagram enable plant-based diets and dieters to *show up differently* and allow for different modes of personal revelation. The chapter concludes by discussing how the capability of vernacular plant-based visual rhetorics to be indistinguishable from the visuality of animal meat contributes to a readiness to partake in plant-based diets; at the same time, this indistinguishability renders plants vulnerable to erasure, which can hamper the transparency needed to get at the challenging questions of what it means to *eat well* (Derrida 1991; Vint 2010).

When aspects of the world are disclosed in unusual ways, a space is created for transformation, and chapter 5 explores how the invitational rhetoric of snake identification communities on Facebook allow snakes to show up differently for people and enable change in human-snakes relations. Participants in these communities learn not only how to identify snakes but they begin to identify *with* snakes from a consubstantial perspective of shared embodiment (cf. Milstein's 2011 work on consubstantiality). Visitors to snake identification boards find that they go from throwing a book when encountering a picture of a snake to playing a game of "can I identify the snake" when images appear on their social media feed. Through sustained engagement with these snake identification communities, people report they are more ecologically ready in their day-to-day lives: able to enjoy their time in the more-than-human world and more likely to intervene to spare the life of a once-feared creature.

Disclosing aspects of the world differently and transforming relations are infrastructural processes for, in the words of Jennifer Clary-Lemon (2019), shortening and thinning the Anthropocene. Chapter 6 introduces the concept of *infrastructuration*—the simultaneous digital, material, and embodied worldbuilding processes of making and being made—to analyze how ideas, materialities, and practices become enabling resources for (de)colonial discourses that afford or constrain (in varying degrees) the mechanisms of the Anthropocene. This chapter also uses *infrastructuration* as a way to illustrate the critical and practical potentials of two worldbuilding projects, the Pink Chicken Project and Reclaim the Void. Ultimately, I argue in this chapter that although both projects participate in raising awareness of anthropocentric, colonial approaches to the more-than-human world, the Pink Chicken Project becomes part of the infrastructure it critiques while Reclaim the Void offers an alternative, affirmative infrastructure for living *otherwise*.

Finally, chapter 7 explores the implications of worldbuilding rhetorics based in a community-care paradigm, particularly the three concepts developed in this book—disclosure, transformation, infrastructuration—for pedagogy-and-practice. Here, I offer that engaging in the work of attending to how digital communities disclose, transform, and participate in building (more-than-)human worlds can contribute to a sense of ecological-readiness, or a willingness to act in, with, and for a (more-than-)human web of relations.

I conclude this introduction the same way I will conclude every chapter in this book—with some observations on vulnerable flourishment born of my daily activities of dwelling on a small rescue farm in the middle of the woods. The reader might experience these stories as a type of posthumanist, pseudo-fable—short accounts of ethical matters told through animal stories. However, rather than the anthropomorphized critters of classic fables, my attempt is to *zoomorphize* human experience through these tales. The stories happen to be true—that is, they are the recounting of events experienced from my subject position—but their facticity or lack thereof is not particularly relevant. What concerns me about these accounts is that they provide psychological and emotional resources for *gathering together* with others in a more-than-human world.

Although these more-than-human beings do not show up in the chapters in a recognizable form of citation—for example "(Red-tailed Hawk 2020)" or "(Emma the Barred Rock Chicken 2021)"—their ways-of-living are both where this journey begins and where the work of discovering what it means to *live well with others* continues after the final period is placed on the last page of this book. The idea of *vulnerable flourishment* came first from living with a flock of chickens and later from engaging with the insights of new materialist, posthumanist, decolonial, and ethics of care scholarship.

STORY TIME: EMMA IS MOLTING

A few months ago on our morning walk, Lawrence and I were talking about our favorite chicken, Emma (named for the English actor Emma Thompson). We were marveling over who she is. Emma got seriously injured when she was a young chicken and an older chicken attacked her. I found her the next morning sitting on top of all the other baby chicks—they were shielded by her body, and she sat on them, seriously injured. Lawrence and I were able to nurse her back to a full recovery—she was a fighter!—though she carries a scar on the back of her head. That experience did not change Emma though, and as we make additions to the flock, she is the first adult chicken to whom we introduce the young chickens. She is always patient with them, even when they are being "a bit big for their britches" and chest bump her. She *does not* let the little ones hurt her, but she corrects them in a beautifully elegant way. I do not mean she does this in a debunked "maternal instinct" fashion (cf. Cooke [2022] on the complicated lives of female nonhuman animals); I mean she does this in Emma fashion.

That's not all! In addition to this gracious gentleness, she carries herself in such a calm, secure manner to this extent: there was once a sharp-shinned hawk (a small hawk) in the backyard that was scaring all the other chickens (including our rooster!), and they ran and hid under the shed. Emma came marching out of that shed going, "Errp errrp errrpp errp errp" and just went about her business. I watched in horror, and then amazement, as the hawk flew away. So, I reflected aloud on all of this that morning as we walked, just utterly amazed at who she is—her kind and gracious strength.

But now, Emma is molting. Emma, our most vibrant chicken, is at her most vulnerable. Normally the boldest of our companion chickens, she backs down to even the youngest and newest of arrivals to the flock. She knows she is in a precarious state: her flight feathers are gone, there is much exposed flesh, and the budding feathers are like pin needles protruding from her skin. So much of her energy is devoted to regrowing these feathers that it is hard for her to forage for nutrients. Walking is a challenge, as she tries to keep the pin feathers from rubbing against her skin. She shies away from getting in the thick of a good forage find, for she does not want to bump into another chicken or get into a standoff. Even targeted attempts to get her extra protein is hard because the mere presence of another chicken—and the potential for a skirmish over the food—causes her to shy away. Emma is, uncharacteristically, hunkering down next to the house until it is time to return to the hen house for the evening. Emma is surviving.

Emma is molting, and Emma is vulnerable. And yet, on the other side of this difficult season—a season in which she can only do her best to take care of herself and live through—she will flourish again. Emma is over five years old, "elderly" for her breed of chicken, and she molts hard twice a year. Right now, Emma cannot experience but only nurture the seeds of future flourishment: strong, beautiful, healthy feathers that will protect her from cold, shield her from rain, and help her to flee from danger.

To the casual observer, Emma's molting may be off-putting. Her naked, pink skin and staggering walk are not "elegant." But Emma is strong and careful, and there is an admirable grace in that. Emma is molting, and Emma is beautiful.

Chapter 2

A Community-Care Paradigm for Ecological Readiness

Addressing the question "What is the good life?" is one of the oldest intellectual pursuits. In Ancient Greece, the question is broached when Socrates considers the centrality of justice to a life well-lived and debates whether the nature of rhetoric contributes to or hinders justice (Plato 2001, 109–21). Aristotle (2011) takes up the question more directly in the *Nicomachean Ethics*, in which he suggests that *the end* of life is *eudaimonia*—that is, the sense of happiness and flourishment that comes from a life well-lived, a life lived in accordance with one's character (11–13). Different theories of a well-lived life have been offered by religions and cultures, but the Anthropocene—a geologic era marked by human impact on Earth's systems and one which raises the question of the livability of the biosphere—necessitates that we attend to wellbeing from a perspective of ecological readiness: to consider what a life well-lived looks like in a more-than-human world. The Anthropocene opens up the existential and ethical question of, *How do we live well with others*, when *others* has to be understood to include nonhuman others? To consider flourishment in the Anthropocene, in this chapter I first introduce some key perspectives on the notion of wellbeing. Next, I offer the community-care paradigm as a means to conceptualize the ontological, epistemological, and ethical considerations of living well in a more-than-human world. Finally, I offer a preliminary, and ever partial, account of what constitutes flourishment in the Anthropocene with all its precarity.

THE ANTHROPO-CENE

Raymond Carver (1989) wrote a short story pondering the eponymous matter of "what we talk about when we talk about love." *Love* is one of our truly difficult-to-define words, like *culture* and *rhetoric*. Another of these elusive terms is *flourishment* and related notions of wellbeing. A difficult term under the best of circumstances, the term is even trickier in the collection of circumstances known as the Anthropocene. The Anthropocene is the name for the era marking anthropogenic—that is, human-induced—effects of earth systems-functioning, including anthropogenic climate change, the physical markers of our synthetic processes and products in the geologic strata, and a species-induced mass extinction event (caused by factors referred to in ecological studies as HIPPO: habitat destruction, invasive species introduced by human trade and travels, pollution, population growth, and overharvesting) (Kolbert 2014). While these consequences of the Anthropocene may be the most remarkable aspects, the human agency-centered rhetorics that treat the more-than-human world as a mere backdrop for human affairs is also of paramount concern for anyone who wishes to see human engagement in the more-than-human world *otherwise*. Here, I am particularly interested in unpacking the "anthropo" of the Anthropocene.

The Anthropocene necessitates consideration of human practices as, fundamentally, ecological practices. The idea of the Anthropocene draws our attention to the ways that human activities in industrial society have ushered in the sixth mass extinction, generated global warming, and placed the biosphere—the very conditions enabling life—in a precarious position. To meet the challenges of the Anthropocene, we have to challenge and re-envision the idea of what it means to be an *anthropo* in a more-than-human world. To begin this work of rethinking the *anthropo*, below, I offer an analysis of anthropocentrism and its affiliated human supremacism and work to disentangle their logics from what it means to be anthropo and, thus, to open new possibilities for understanding, challenging, and changing the Anthropocene.

Anthropocentrism, in its most broad sense, refers to a human-centered perspective. In philosophy, this term often refers to a perspective that humans are somewhere between "only" and "primary" holders of moral standing. For me, this perspective shows up in daily life whenever there is a natural or technological disaster reported on the news and someone says that no lives were lost. While I am always thankful to hear of no human fatalities, I am always skeptical that *no* lives were lost in these events. By drawing attention to this example, I do not mean to imply that the news anchors or news writers are callous in their reporting; rather, it is in perfect accordance with an anthropocentric logic not to consider more-than-human lives when humans are considered the only, or at least primary, holders of moral standing. In anthropocentric value systems, the more-than-human world shows up as a *standing reserve* for human activity (Heidegger 1977, 17–19). I will return to this idea of the more-than-human world existing *for* humans later in the chapter when I discuss the community-care paradigm, but for now it is sufficient to register that anthropocentrism places one species—humans (and often, a very particular, limiting, and Eurocentric understanding of what it means to be human)—as the organizing principle for the rest of the planet and its life-forms. Anthropocentrism is closely related to the idea of *human supremacism*. *Human supremacism*, founded on anthropocentric logic, is a belief that humans are *entitled to* act on the more-than-human world in whatever way is chosen, even if it means sacrificing "the continuation of life on the planet" (Jensen 2016, 17).

Although I share Jensen's concern about human entitlement toward the more-than-human world, my concern differs in two respects. One, I believe that human ways-of-life are still in relation and subject to change (more on this later in this chapter), and two, I doubt that many humans are intentionally choosing to view the world through a lens of human supremacism. Rather, through the anthropological machine, anthropocentric perspectives and even human supremacism show up as a "natural" state of affairs. The *anthropological machine* (Agamben 2004, 37–38)—consisting of institutions, systems, practices, and philosophies from ancient through to contemporary life—is a "*performative* apparatus, inasmuch as it enacts and calls into being (which is to say, performs) a certain reality. It is the machine itself that creates, reproduces, and maintains the distinction between human and animal life" (Calarco 2014, 54). Agamben's notion of the *anthropological machine* is a rich concept with which one could productively think about the Anthropocene much more deeply, but for the purposes here, it is sufficient to register that there are many component parts in everyday Western life that perform the distinction between human and nonhuman animal and make it hard for the more-than-human world to show up otherwise—or even show up at all. Indeed, animals are so pushed to the edges of consciousness that although children frequently dream of animals, adults rarely do (Bugalho and Paiva 2011).

One could say this is splitting hairs; that it matters little whether anthropocentrism is chosen or thrust upon people; what matters is that it *is* the state of affairs, that the "anthropological machine" continues to be a well-oiled one. Certainly, the Anthropocene would seem to support such an anthropocentric understanding of how things work: the era is named as such to mark human's ability to alter the functioning of earth's systems. However, there are some problems with the term *Anthropocene* that must be noted. While the term appropriately draws attention to human agency and consequence, it simultaneously obscures other aspects of the present moment in geologic history.

I find hope in the idea that anthropocentrism might be part of an anthropological machine because machines can be monkeywrenched (Abbey 2006), taken apart, reassembled, rethought, and remade into something new. The potential of the anthropological machine may be immense but it is not absolute, and I hold open that we do not even know what an *anthropo* can do or be. From species agnosticism[1] to relational ontologies (described in the next section), there are reasons to believe that the anthropological machine is only one possible performative apparatus among many, at least three of which include critiques of the Anthropocene, the more-than-humanness of the Anthropocene, and debates about human nature.

First, there are the critiques of the Anthropocene that are most vividly marked by the other names suggested to mark the era, like the Capitalocene (emphasizing the role of capitalism) or Plantationocene (emphasizing plantation-logics that discipline multispecies into forced labor). Donna Haraway suggests that the term Anthropocene is too sweeping, that it ignores the socio-ecological, historic-situatedness of the human activities that have brought about the Anthropocene. The effects of the Anthropocene are effects of certain activities of certain human lifestyles and are "not humankind all the time everywhere" (Haraway 2018, 80). Further, the Anthropocene does not make salient the multispecies relations necessitated to enable industrial human activity. The Anthropocene not only affects the more-than-human world but also would not be possible without the matrix of multispecies relations—including the disciplining of plant and animal bodies in industrial agriculture—that afford the industrial, consumerist infrastructures and lifestyles productive of the Anthropocene (Haraway and Tsing 2019).

The more-than-human conditions and consequences of the Anthropocene bring me to another critique of what the Anthropocene conceals: how it actually exceeds humanity. The Anthropocene, as a name, may serve to overinflate humanity's grasp of the situation. Although the Anthropocene may be marked by human activity—for example, the amount of carbon dioxide pumped into the atmosphere as a side effect of industrialization, or the deforestation occurring to meet the increasing demand for meat by an ever-growing human population—these activities and their networked interactions with the more-than-human world *exceed* humanity. They are, as Timothy Morton (2013) describes, hyperobjects, or phenomena of such spatiotemporal vastness that they exceed human grasp. This observation may sound simple but it is quite profound because the Anthropocene calls on humans to act toward the probable good in situations with vast unknowns. Poet Nikki Giovanni (2021) described in an interview imagining being visited by extraterrestrials and what she would say to them if they asked her what she was. She said the only thing she could think of was Earthling. "Black" and "woman" would likely have no meaning to these intergalactic travelers.

While the term Anthropocene is only meant to mark a time in Earth history, it can invite a sense of Anthropocentrism not only on Earth but beyond the realm of Earth's gravitational pull, as well. In short, it can make it difficult to sustain the sort of humility necessary for the work of living differently.

Third, and harkening back to Haraway's "not humankind all the time everywhere," it is important to ask *what even is an anthropo*? The question of what constitutes "human nature" is one that has been asked throughout the ages and across cultures. Not only is there no consensus around the answers to this question, but what answers are supplied tend not to hold up for long. My objective in asking this question is not to say there is no such thing as human. Indeed, it would be defeating the premise of this book that the Anthropocene requires *Homo sapiens* to dwell differently with fellow Earthlings. Instead, I am saying that what humans can do is to remain open. In *The Dawn of Everything*, anthropologists David Graeber and David Wengrow (2021) contend that what most accounts of human history get wrong is the search for some sort of determination or cause in the first place—whether it be economic and human-generated or geographic and therefore more-than-human. Rather, they argue, humans are not fundamentally anything. Humans are playful creatures who interact with things—ideas, landscapes, other animals, plants, languages, political arrangements, economic systems—and there are many ways of being human. The travesty for Graeber and Wengrow is that modern humans seem to have forgotten this realm of possibility and become locked into a narrow conception of what it means to perform a human way-of-life. Thus, the kind of anthropocentric thinking that leads to a sense of *anthropodeterminism,* that human nature is somehow determined and in-turn determining what occurs in the more-than-human world, are not only inflated and detrimental to the more-than-human world, but they are also limiting and inherently colonistic to the variety of ways-of-life that make up the ever-expanding mosaic of humanity.

Of course, the causes and critiques of the Anthropocene exceed what is captured above. What is important at this point in the journey is to recognize both how the Anthropocene calls humans to act (differently) *and* that what it means to be an *anthropo* in the Anthropocene remains in play, remains a matter of invention. Presently, the Anthropocene is characterized by the devastating consequences of human activity on the lives of those deemed less-than-human. However, this book is dedicated to the core assumption that when one thinks about the possibilities for enacting *anthropo*, it is better to think Silly Putty than cement. There are many ways to live as an anthropo in the Anthropocene, and some of those ways-of-life are more conducive to living in more-than-human communities than others.

A COMMUNITY-CARE PARADIGM

Thinking through what flourishing means and how it happens in the Anthropocene, I find myself turning to a *community-care paradigm* for reflecting on human discourse and activity in the more-than-human world. Before conceptualizing what it means to flourish in a time of great vulnerability, I think it is important to provide something of a topographic map of this paradigm (something like a street-level view would require its own book) by focusing on its ontological, epistemic, and ethical dimensions, as well as the smallest unit of analysis to consider. I want to be *care*ful—that is, full of care—in how I discuss a community-care paradigm.

This paradigm aligns with the ecological frameworks and ethical engagements of Calarco's (2020) concept of indistinction, Gruen's (2015) entangled empathy, Kimmerer's (2013) grammar of animacy, Riley's (2014) democracy of creatures, Curry's (2011) green citizenry, van Dooren's (2018) conceptualization of extinction, Giraud's (2019) ethics of exclusion, de la Bellacasa's (2017) speculative ethics of care, Clary-Lemon's (2019) new materialist environmental rhetoric, Barnett's (2019) earthly coexistence, Quartz's (2022) earthly *eudaimonia*, and the Indigenous cosmovision (Kemmerer 2017) and perspective of "being relations" (Chilisa 2019, 110). The preceding list, as long as it is, actually names but a few concepts that support community-care *ethico-onto-epistemologies* (Barad 2007, 185) and foreground the deep entanglement of the more-than-human world in human affairs and vice versa. The community-care paradigm I offer in no way supplants these ideas, nor does it equate them. Each of these perspectives offers unique insight that is neither identical nor reducible to another.[2] I see the community-care paradigm as made possible by and in concert with these more-than-human frameworks and concepts. To paraphrase Donna Haraway, the community-care paradigm works in addition, not subtraction (Haraway and Tsing 2019).

Beyond these concepts, I conceptualize the community-care paradigm as taking into account two other articulations of human relations in and with the more-than-human world. One is Phaedra Pezzullo's and J. Robert Cox's (2021) description of environmental communication as a discipline of crisis and a discipline of care. As a discipline of crisis, environmental communication—and I would imagine most of what constitutes the environmental humanities—is crisis-oriented and dedicated to the study and practice of addressing environmental threats. As a discipline of care, environmental communication recognizes interconnection and interdependence, and values biodiversity and interspecies relations.

One other articulation of human affairs in the more-than-human world considers Aldo Leopold's (1987) famous quote in the forward to *A Sand County Almanac*: "We abuse land because we regard it as a commodity belonging to us. Where we see land as a community to which we belong, we may begin to use it with love and respect" (viii). As J. Baird Callicott (2017) elaborates, this ecological community has both holistic and individual registers and is neither "inhumane nor inhuman" (259). Humans are not asked, for example, to allow a tick to consume the creature's fill of one's blood, but humans are asked to consider how to care for the more-than-human world in one's efforts to exclude ticks from one's body. Permethrin, a common spray used to treat clothes and yards, introduces toxicity into an area—excluding ticks through an exterminating mechanism that is also toxic for bees and fish (through rainwater runoff) (Environmental Protection Agency 2006).

Putting crisis and care into conversation with commodity and community enables me to generate a matrix of paradigmatic positions a person can hold toward the more-than-human world that offer alternatives to the prominent positions of humans as saving or consuming the more-than-human world. This matrix is not exhaustive, and the boundaries are not firm. Indeed, in line with Pezzullo and Cox, I would suggest that the community-care paradigm must have a porous boundary with the community crisis paradigm to recognize and meet the robust challenges of the Anthropocene. To engage in community-care work without attention to crisis is a mistake that runs the risk of perpetuating *toxic positivity* discourse and behaviors. However, I want to focus *this* work on the community-care paradigm because there already exists an abundance of publications (e.g., *This Is the Way the World Ends, The Madhouse Effect, The Uninhabitable Earth*) that speak to a community-crisis perspective and do the important work of highlighting the dangers posed by the Anthropocene.

The community-care paradigm, at its broadest and most basic, functions to acknowledge that humans are embedded in, part of, and actors with deeply entangled more-than-human world relations. Again, this paradigm is not wholly distinct from the community-crisis paradigm but sees interconnection, interdependence, biodiversity, and interspecies relations as resources for rhetorical invention (discussed more in chapter 3) for humans living otherwise in their more-than-human communities. To illuminate the community-care paradigm more fully, I continue by discussing its ontological foundations, its ethical commitments, and pointing to what I call *individuals-in-communities* as the basic unit of analysis in a community-care paradigm. Table 2.1 shows how a community-care paradigm compares with other paradigmatic models of human positionality in the more-than-human world.

Table 2.1. Commodity/Community-Crisis/Care Model

	Commodity	Community
Crisis	*Commodity-Crisis Paradigm*: Nonhuman world viewed as a standing-reserve for humans (Heidegger 1977) *Interpretation of Commodity-Crisis*: Crisis is a crisis of production and likely to be understood as necessitating greater human ingenuity and technoscientific progress to overcome shortcomings (Shellenberger and Nordhaus 2011)	*Community-Crisis Paradigm:* Humans are part of the world, *and* the more-than-human world is suffering from human parasitism (Pauly 2014) *Interpretation of Community-Crisis:* Soft interpretation: Human activity is at odds with the greater ecological interest (Jensen 2016); Hard interpretation: The more-than-human world may be better off as a nonhuman world (Voluntary Human Extinction Movement 2022)
Care	*Commodity-Care Paradigm:* Nonhuman world needs good stewardship practices to continue to provide resources (Riley 2014 on stewardship versus democracy) *Interpretation of Commodity-Care:* Efforts to conserve resources and engage in sustainable practices to ensure future availability (Valladolid and Apffel-Marglin 2001 on sustainability)	*Community-Care Paradigm:* Humans are part of the world, and all entities making up the more-than-human world have intrinsic worth and claims to flourishment (Calarco 2020) *Interpretation of Community-Care:* The more-than-human world is a democracy of subjects and necessitates a green citizenry (Curry 2011)

Table created by author.

ONTOLOGICAL FOUNDATIONS: RELATIONS ALL THE WAY DOWN

Many relational ontologies abound, from Indigenous cosmovisions to Karen Barad's (2007) agential realism to science and technology studies' actor-network theory to evolutionary biology, which takes seriously the role of the environment in shaping species and individuals. While there are important distinctions within each framework of relational ontology, there are some broad, overlapping themes that can be drawn out: change is constant; all phenomena (from discourse to action to material bodies) occur in continuous gradients; phenomena are contingent; it is more productive to investigate verbs (processes) than nouns (entities); things—as we understand them—are generated by practice and performance and not from a prior essence; thought and the world are interwoven; and phenomena exist in the multiple. I am less interested in crowning *a* particular relational ontology *the* relational ontology (which would work against the basic tenets of a relational ontology in the first place) than in drawing attention to the vibrant threads that run through these ontological perspectives.

Before attending to these threads, it is worth considering what a discussion of a paradigm's ontological underpinnings offers. Martha Nussbaum (2006a) argues persuasively that actions of justice (and care) are far less dependent on establishing a particular ontologic framework than a robust political framework. She asserts that multiple ontologies can still converge on good actions, and her emphasis is on how we get to those good actions. I agree with Nussbaum, and I believe that one could arrive at *a* community-care paradigm that is grounded in a theological ontology or an object-oriented ontology. (Although, I think saying "other ontological positions could also yield different insight on this phenomenon" is in itself a highly relational ontological statement to make!) However, these would be *different* community-care paradigms—not to say better or worse, but different. With that caveat, I am now able to elaborate on how a community-care paradigm enacts a relational ontology.

Relational ontologies ground a community-care paradigm in a posthumanistic or alternative-to-humanistic stance (e.g., Indigenous cosmovisions) that can contribute to an ecologic ethic of care (to be discussed shortly). A posthumanist or alternative to humanist stance is not an *anti*-humanist stance aimed at shaming humans or advocating for human extinction. Rather, posthumanism and alternatives to humanism offer deconstructions of Enlightenment conceptions of the atomized individual. Indeed, quantum physics challenges the atomized notion of atoms—given their quarks and other subatomic particles!

What the deconstruction of the atomized individual entails is that individuals have to be understood not as profoundly individual subjects who can "pull themselves up by their bootstraps," or to put it more biologically, the Richard Dawkins notion of the selfish gene in which it is truly an individual fight for survival. Rather, individuals are understood as profoundly relational and open to being *otherwise*. This idea is captured well by Anna Tsing's (2015) observation about the world(re)making possibilities of transformative encounters: "Without the possibility of transformative encounters, mathematics can replace natural history and ethnography" (28). What makes these transformative encounters world-making and not only subject-making is that the vibrations of transformative encounters extend in all directions because they occur in a "web of mutual transformations and deeply meaningful encounters that are always embodied" (Weber 2022).

A relational ontology has dramatic impact on notions of agency and ethics. Agency—as a relational, more-than-human concept—is separated from intention. Certainly, intention is still a primary component for many instances of agency, both human and nonhuman. When I go to the grocery store or a hawk grabs one of my chickens, both of us are compelled by an intention to feed. However, when the bars of the porch allow the chickens to pass through to safety and prevent the hawk from entering, the porch assemblage—of wood, nails, spaces of openness and closedness—has certainly shown up as an actant in that encounter. Another example of how agency exceeds intention is illustrated by Maria Puig de la Bellacasa (2017), who describes how earthworms participate in caring for the soil. Their activities—seemingly without any intention to do so—help make soil rich for growing, allowing the flourishing of the plants, which then allows the flourishing of all fauna (218). De la Bellacasa points out that earthworms care for the soil but do so in ways that invite a reconsideration of what it means *to care*. De la Bellacasa thinks with earthworms to recast Tronto's notion of caring as "a [more-than-human] species activity that includes everything we do to maintain, continue, and repair our 'world' so that we can live in it as well as possible. That includes our bodies, our selves, and our environment, all of which we seek to interweave in a complex, life-sustaining web" (Tronto 1993, 103), and with understanding that the "we" and "species activity" far exceed humans and human intentionality.

Distributed agency is central to the community-care paradigm I am describing, and it is important to emphasize both of those terms in turn. *Distributed* agency announces that humans exist in a dense network of actors/actants—some human, some more-than-human flora and fauna, some more-than-human digital technologies, and so on—all constituting communities that afford different degrees of care for the individuals in these communities. Distributed *agency* emphasizes that, even in a dense network of actors/actants, humans still have the capacity to act, and those actions are responsible to a host of sociopolitical and ethical claims made on human actors.

A relational ontology decenters the Enlightenment subject in human actions. Actions grounded in a persistent commitment to the Enlightenment subject miss the mark of relational transformation that occurs in communities: "Such a schema based on sameness to the human treats earth others as of value just to the extent that they resemble the human as hegemonic centre, rather than as an independent centre with potential needs, excellences and claims to flourish of their own" (Plumwood 2002, 167). Instead of an *a priori* understanding of what a human is, something I have already described as being a limiting idea to address the challenges of the Anthropocene, a relational ontology attends to what it means to act when difference is given *and*, despite the difference, we are all wrapped up in a web of relations that are mutually—though not identically—affective.

In addition to relational ontologies, there are *indigenist* relational cosmovisions—the physical and spiritual worldviews of Indigenous communities. Here, the term *indigenist* is used to distinguish Indigenous ideologies shared by Indigenous and non-Indigenous peoples—that is, a political ideology of solidarity with the claims of Indigenous people—from *Indigenous ideology*, which could connote Indigenous authorship (Hidalgo-Capitán, Cubillo-Guevara, and Masabalín-Caisaguano 2020).[3] Indigenist worldviews embrace a relational and performative ontology that is opposed to the individualism of many Western ontologies and understands material, social, and spiritual realities to be a collective experience in the context of place (Haverkamp 2021; Yates 2021). An example of how this cosmovision shows up in language can be found in the Māori (an Indigenous Polynesia community in New Zealand) word *whenua,* which is the word for *placenta* and for *land. Whenua* articulates both placenta and land as providing support, nourishment, security, and anchorage (Wilson, Boulton, and Warbrick 2019, 71). What might the current geologic era look like if a worldview of such interconnectedness were more widely shared?

A community-care paradigm, however, cannot stop at a relational ontology because, as Eva Haifa Giraud (2019) notes, nothing may exist outside relations but not all relations are relations of flourishment. Beyond a relational ontology, people still have to make decisions and *act*, as best as possible (173). A commitment to flourishment in the Anthropocene is necessarily a communal affair, and that community is more than human. A commitment to flourishment—to caring for our communities—in the Anthropocene has to be partial, open, response-able/responsible, and contestable, but that commitment still requires acting in the world individually and always with others. None of the actions are innocent; instead, what is possible is an intention to act as well as possible in the world. Acting "as well as possible" involves epistemic and ethical concerns: how does one know what it is to act well in, and what does it mean to care for, a more-than-human world? Thus, I now turn my attention to matters of knowing and acting in a more-than-human world.

EPISTEMIC CONSIDERATIONS: KNOWING THE MORE-THAN-HUMAN

The next consideration in the community-care paradigm is what it means to *know* something in and about a more-than-human world. Like the discussions of community-care ontology, ethics, and unit of analysis, the discussion of what counts as knowledge and how knowledge gets produced could be a manuscript-length treatment. However, for my present purposes of orienting the reader to the community-care paradigm and sketching its form without over-stipulating its characteristics, I limit this discussion to introducing the epistemic position I see as most-suited to a community-care paradigm, and briefly discuss what this means for knowing the more-than-human world.

The *situated knowledges* approach of feminist science studies is well-suited to the community-care paradigm I am proposing because its core principles carry an ethos of epistemic humility that is a necessary counter to the hubris of the Anthropocene. *Situated knowledges* refers to knowledges based on embodiment and embeddedness in the historical, cultural, linguistic, and value contexts of the knower, in which neither experiencing-subjects nor "nature" can be treated as straightforward entities (Haraway 1991, 109). The logic of *situated knowledges* works against an epistemic position of absolute, universal knowledge—or what Donna Haraway refers to as the god trick (Haraway 1988, 581). Rather, *situated knowledges* affirms that *knowing* is not a Platonic kind of transcendent knowing; knowing is better understood as an embodied affair that is entangled with an individual's personal experiences, background beliefs, cognitive styles, and knowledge-relations. This knottiness of situated knowledges is perhaps best expressed in Karen Barad's (2007) articulation of *ethico-onto-epistem-ology*. The notion of ethico-onto-epistem-ology gets at the inseparability of ethics, ontology, and epistemology in any act of knowledge production.

Situated knowledges, sometimes referred to as epistemic relativism, can be compared to *doxa*. *Doxa*, from the ancient Greek, refers to the realm of ambiguity, becoming, opinion, and multiple perspectives (Jasinski 2001, 183). In the Western tradition, the pluralistic approach to knowledge recognized by *situated knowledges* and *doxa* has been critiqued since Plato. Critics contend that such epistemological positions work to undermine the very science that provides information on the causes and consequences of the Anthropocene; indeed, critics argue that the very concept of knowledge in undermined. In a moment, I will address this concern, for it is an important one, but first, I want to open up this critique a bit.

There are different senses of what it means *to know*. There is *knowing-about-x* (knowledge of an entity), *knowing-how* (knowledge of processes, practices—how to do things), and *knowing-that* (knowledge of facts). I doubt that even the most skeptical of *situated knowledges* as an epistemic position would doubt the embodied and relational qualities in knowing-about-x or in knowing-how. What I think is really being debated are claims to knowing-that, what it means to produce a fact. Returning to the concern about situated knowledges as epistemic relativism, the concern is whether *situated knowledges* undermines the knowing-that—the knowledge and production of facts—of scientific claims.

The short answer is no. Claiming that knowledge is and must be locatable and is always tied to ontological and ethical concerns is not the same as doubting that scientific processes are ways of *knowing-that*. As Robert Hariman (1986) points out, humans are limited to a *doxastic* world. *Complete disclosure* of an entity is impossible because every way of knowing acts as a filter and because entities are always in a state of becoming. However, to say that complete disclosure—a final and settled Truth—is beyond human grasp is *not* to say the methods of science do not disclose important information about entities. In Haraway's foundational 1988 article, she was careful to articulate that a situated understanding of knowledge does not negate that good apparatuses of knowledge exist in the world. The key is not rejecting all apparatuses of knowledge because they are partial and contingent; the key is exploring how the apparatuses of knowledge produce knowledge, how this knowledge is situated in cultural and historical contexts, and how this knowledge is entangled in ontological and ethical matters. Every apparatus of knowledge provides a way of knowing which is simultaneously a way of not knowing (Burke 1966, 29). This feature of simultaneous illumination and obfuscation is not a flaw to be corrected but an inherent feature to be appreciated.

Perhaps a brief illustration here will help clarify the ethos of situated knowledges. Anthropologist Eduardo Kohn (2013) writes that, "People in Ávila take great pleasure in finding a viewpoint that encompasses multiple perspectives" (97), and offers an Ávila myth as an example. The story involves a hero who is patching his roof. He calls out to a man-eating jaguar on the ground, addressing him as "son-in-law," to help him by going inside the thatch and poking sticks through the holes in the roof. Patching the roof is a two-person task. The person on the inside can see the holes because the sunlight shines through them, but the person on the inside cannot patch the holes because the roof is too high. The person on the roof can patch the holes but cannot see them. With the help of the person inside, the inside and outside perspectives become uniquely aligned and afford a special ability to be acted on. The jaguar—despite being a person-eating jaguar—assists the protagonist because the jaguar has been hailed as son-in-law and feels compelled to fulfill this role.[4]

Kohn's recounting of the Ávila myth speaks to the synergistic potentials of situated knowledges. Instead of assuming one universal knowledge position, a person can be open to aligning knowledges. Kohn pushes this alignment even further, arguing that the coming together of these divergent perspectives "makes available something about life 'itself'" (98). Also important in the Ávila myth is the indistinction of human and nonhuman knowledges in terms of hierarchy. The human on the roof had different but not superior knowledge to the jaguar, and the reparative work was accomplished through multispecies, multiperspective knowledges. The Ávila myth raises the issue of what it means to *know with* the more-than-human world.

One recurrent problem in more-than-human studies is the unidirectional nature of much of the research. Humans often try to figure out how much nonhuman animals know, including how many human words a nonhuman animal understands, and then make claims about the relative intelligence of that species. Although such studies aim to unseat human supremacism by raising awareness of the capacities of other species, the discussions are frequently framed in ways that maintain human intelligence as the measuring stick. Rather than trying to figure out how many human words a pig knows, we could flip the script and ask, "How many words of pig do you know?" (Jensen 2016, 29).

Similarly, primatologist Frans de Waal (2016) provocatively asks, "Are we smart enough to know how smart animals are?" Throughout de Waal's book, there are amazing examples of knowledges that are impressive and not-at-all human, and some humbling accounts of anthropocentric, epistemic hubris. I do not want to go too far down the rabbit hole of more-than-human cognitive capacities here. Rather, I simply want to mark that, as I move into a discussion about what it means to have knowledge of the more-than-human world, a situated-knowledges approach affirms that the more-than-human world has knowledges that constitute this shared space called Earth.

Sharing Earth with more-than-human others raises challenging questions of what it means to *know* a fact about these others. Often, humans find that something they *knew* about a nonhuman animal was inaccurate. For example, one of those accounts of anthropocentric, epistemic hubris I mentioned a moment ago concerns how elephants initially "failed" the mirror test. The mirror test refers to whether nonhuman animals recognize their reflections in the mirror, and this is thought to be related to a sense of self. The mirror test is interesting; it is interesting to consider what creatures have life-worlds and ways-of-life that are generative of the perceptual capacities to attend to one's reflection. The mirror test as a lone measure of sense of self is highly problematic: It can display if an animal has the particular sense of self that is captured by perceptual recognition of one's reflection, but it certainly does not provide evidence of the absence of a sense of self. Further, there are issues in how the test gets administered. Elephants initially failed the mirror test until it was pointed out that the mirrors' angle and placement may have been at issue. The mirrors had been placed in such a way that likely all they saw in the mirror were their legs behind the enclosure's bars. When the mirrors were raised to meet the elephants' gaze, they met the criteria of the test.

The preceding example is not meant to chastise the researchers who placed the mirrors too low for the elephants. It is exceedingly difficult to try to understand and design for another species' perceptual capacities. The example is meant to illuminate (1) how a posture of situated knowledges—that acknowledges that all knowledge is partial, incomplete, impure, and subject to revision—accounts for the knowledge-making practices, and (2) what is at stake for the more-than-human world in our knowledge-making practices (i.e., ethico-onto-epistem-ology). In November 2021, lobsters were declared sentient in the UK, meaning they were now subject to the UK's Animal Welfare Bill. With the recognition that lobsters feel pain, methods of slaughter need to be revised (Aridi 2021). In chapter 4, I discuss matters of eating well and what makes a being killable, but for now, it is sufficient to note that matters of *knowing-that* are not only subject to change but are intimately associated with authorizations to act on others.[5] Knowledge is not neatly cordoned off from other dimensions of living with, and that is why taking account of the apparatuses of knowledge-production is a crucial aspect of the community-care paradigm. At the same time, I concur with Nussbaum (2022) that knowledge can "help us to think better about ethical questions before us" (34) and develop better theoretical orientations that can direct law and policy. The community-care paradigm recognizes that knowledge is partial and yet humans still have to act in the world with the knowledge-at-hand.

Thus far, my epistemic accounting has focused on the necessity for, yet provisional nature of, knowledge for action. Now, it is time to offer some sense of how this necessary and provisional knowledge gets produced in a situated, more-than-human world. There are a number of approaches taken to generating knowledge of the more-than-human world: sympathetic imagining (Nussbaum 2006a), Indigenous epistemologies (Seraphin 2020), and critical anthropomorphism (Burghardt 2016), to name a few.

Although there are important distinctions among these generative frameworks, which will be explored further momentarily, they have in common that the most parsimonious explanations of more-than-human lives are *continuous with* rather than *separate from* human lives. This is not to ignore the differences within and among species. Instead, it is an acknowledgment that human bodies are animal bodies, and especially when nonhuman animals are being studied, some degree of continuity among embodied experience would be expected based on the presence of universals, homologies, or analogies (Marchesini 2015, 78).

Universals are those things that can be said to be applicable across species. For example, animals repeat behaviors that bring pleasure and move away from sources of pain. Next, there are *homologies*, characteristics found among closely related animals due to a shared ancestor. For example, I can reasonably interpret another primate—with whom I share a recent ancestor on the evolutionary timescale—refusing to complete requested tasks after a conspecific (i.e., a member of the same species) receives a better reward for completing the same task due to feeling something like being treated unfairly (Brosnan and de Waal 2003). Finally, there are *analogies*, shared characteristics developed through similar environmental pressures. For example, given humans' long co-evolution with our canine companion species, many dogs have the capacity to follow our pointing, learning that the gesture serves as a referential cue. Thus, when one hears a dog yelp in response to being bitten by another dog, the most parsimonious explanation is that the yelping dog is in pain. To offer an alternative explanation would require offering an evolutionary accounting for how a yelp in humans developed so distinctly from a yelp in dogs.

To put the expectation of continuity in sharper relief, I offer an anatomical example. Whenever a previously undocumented species of animal gets recognized in the scientific community, the researchers are not tasked with demonstrating that the cephalic orbs are eyes. It may be unknown what kind of vision the animal has, but it is accepted that these cephalic orbs are part of a visual perceptual capacity and not, say, how the being takes in nutrients. If somehow these orbs were how the being took in nutrients, there would be a great deal of work involved in explaining the evolutionary processes and pressures that led this near-relative of primates, rodents, and cetaceans to have evolved so differently—especially when the visual function of these cephalic orbs is shared by such distant relatives as the octopus. Looking for an explanation that would separate humans *despite* the presence of evolutionary universals, homologies, or analogies, would seem to be an example of *anthropodenial*, the willful ignorance of shared characteristics among humans and other animals (de Waal 1999).

Sympathetic imagining, Indigenous epistemologies, and critical anthropomorphism are all ways of generating hypotheses about the more-than-human world that expect continuity among life-forms. Sympathetic imagining and critical anthropomorphism use the sentience of the observer to consider the lives of nonhuman others. Sympathetic imagining places sympathy as crucial to knowledge-building as sympathy is the faculty that allows people to share at times the being of another in fallible, speculative ways (Nussbaum 2006a, 355). Critical anthropomorphism acknowledges that there are both similar and different attributes among humans and other animals but seeks to identify a point of comparison in order to come to know these attributes because "it is more correct to identify that term [of comparison] in the human being rather than in a form of machinery" (Marchesini 2015, 79). Indigenous epistemologies—a problematically broad term that both gestures to de/anti-colonial ways of knowing *and* reduces vast distinctions under an enormous umbrella—attend to embodiment and embeddedness in a more-than-human world and evoke an understanding of epistemology as together-doing knowledges (i.e., making, creating, thinking) that is necessarily always already more-than-human (Forbes 2001).

Knowledge about the more-than-human world is not so much knowledge that speaks to a fundamental similarity or difference between one species and another but about the ongoing, dynamic quality of different lifeways. An important thing to mark in this perspective is that the human is not some immutable reference point from which to assess other capacities; instead, the emphasis is on the *ways-of-life*, some of which are shared in interesting and often surprising ways and others of which are unique. A coming-to-know something is a coming-to-know of a process in relation, not a coming to know of a static entity.

The distinctions among sympathetic imagining, critical anthropomorphism, and Indigenous epistemologies are large and important. However, what I want to preserve in this account of a community-care paradigm is an epistemological humility that roots knowledge in the more-than-human world. Despite their distinctions, each of these frameworks embraces humility and understands the observer as a fully embodied individual. Each can provide partial and provisional knowledge that aids humans in acting well in a more-than-human world, and it is to the work of acting well in a more-than-human world that I now turn.

ETHICAL FOUNDATIONS: CARE

So far, I have discussed the context of the Anthropocene and how a relational ontology and situated epistemologies can help us to rethink agency and ethics in the Anthropocene, but I have stayed pretty zoomed out, discussing these matters rather abstractly. Now, I want to turn to considerations of what the ethics of encounters looks like in the Anthropocene, to focus on what it means to live well with others. Perhaps one of the most elegant descriptions of *good* and *bad* encounters is offered by Benedict de Spinoza, carefully rearticulated and accentuated by Gilles Deleuze.

For Spinoza, encounters are *good* or *bad* depending on how they increase or decrease one's—with the understanding that *one* is itself a relationally composed entity—power to act, one's power of living (Deleuze 1988, 72). As noted in chapter 1, the paradigmatic example of bad relations in Spinozan philosophy is the interaction between a poison and the body. A poison decomposes the body. Conversely, remedies restore the body. "Therefore, everything that is bad is measured by a decrease of the power of acting (sadness-hatred); everything that is good, by an increase of this same power (joy-love)," Deleuze writes (72). While I do not mean to oversimplify the distinctions between Spinoza's ethics and *eudaimonism*, it is worth taking a moment to point to some overlap. Deleuze writes that according to Spinoza, "It is a disgrace to seek the internal essence of man in his bad extrinsic encounters. Everything that involves sadness serves tyranny and oppression" (72). With this quote and the one preceding, we can see some connection in both Spinoza's ethics and *eudaimonism* in that joyful fulfillment is considered in a context of good extrinsic encounters, good relations.

Now, Spinoza (1996) himself was anthropocentric in *Ethics*, writing that "the lower animals have sensations. But I do deny that we are therefore not permitted to consider our own advantage, use them at our pleasure, and treat them as is most convenient to us. For they do not agree in nature with us, and their affects are different in nature from human affects" (135).

However, Spinoza's ethics of encounters does not necessitate anthropocentrism. As Deleuze (1988) expands on Spinoza, we see that "relations compound without limit" and that "what is good is any increase of the power of acting" (72) and, further, that this power of acting "is what opens the capacity for being affected to the greatest number of things" (71). While Spinoza may have been anthropocentric in his articulation of ethical encounters, there is nothing that necessitates anthropocentrism in these ethical frameworks. Indeed, understanding the more-than-human world as involved in human relations, as Deleuze does, seems to recognize more appropriately the capacity for being affected by the greatest number of things.

To my reading, the ethical framework that best attends to what it means to be affect(ed)(ing)(able) by the greatest number of entities and processes is an ethics of care. Bringing care together with Spinozan ethics, *care* is the attention—both internally via thoughts and feelings and externally via behaviors—to others' potentials, and it is through the maximization of others' potential that each individual is opened to the greatest range of experiences and knowledge—that is, to the greatest possible flourishment.

The concept of ethics of care has historical roots in psychologist Carol Gilligan's (1993) refutation of research that claimed women lacked the moral development of men. Gilligan sought to demonstrate that women's morality developed differently. Where men were concerned with ethics of justice that emphasized rules and regulations, women were concerned with ethics of care that revolved around relationships and responsibilities. Gilligan is said to have started a movement, and like many movements, there have been waves.

Gilligan's work is associated with the first wave of feminism and has been critiqued as (1) being Eurocentric because the sex differences Gilligan found did not exist in the African American community, (2) reifying sex differences that ignore the performance of gender and could potentially—harmfully—establish women as innately moral agents, and (3) perpetuating an oppressive position of women as natural, and therefore rightful, caregivers (Card 1990, 101; Collins 2000, 8). Building on Gilligan's work and its critiques, Joan Tronto (1993) situates care as a practice and disposition of caring about, taking caring of, caregiving, and care-receiving that includes four ethical criteria: attentiveness (e.g., the recognition of needs); responsibility (e.g., evaluating the efficacy and ethicality of how one engages in acts of care); competence (e.g., concern with the consequences of care); and responsiveness (e.g., remaining alert to the possibilities of abuse in caring relations).

More recently, Maria Puig de la Bellacasa (2017) built on Tronto's notion of the active work of care and reconceptualized care beyond an anthropocentric lens of intentional acts and articulates care as the labors of beings in their relations and environments. She writes that care is not about normative obligations but rather about thick, impure involvement in the world. We are always already involved with more-than-human life forms, forces, and objects. De la Bellacasa's insights are consonant with indigenist approaches to ethics of care that emphasize the interconnected work of living well with others where *others* include other humans, other beings, and Earth (Haverkamp 2021; Yates 2021). Part of a feminist and indigenist ethic of care is about enacting "living in difference together," but it also involves *learning* how to live in difference together. In this paradigm, learning is understood as a thinking-feeling praxis (Haverkamp 2021) of knowledge. Nussbaum (2006b)—operating outside the ethics of care tradition but who is concerned with issues of justice for the more-than-human world—argues for the necessity of generating knowledge when it comes to matters of care because of the malleable and corruptible nature of sympathies. "That is why we typically need philosophy and its theories of justice. Theories help us to get the best of our own ethical intuitions, preventing self-serving distortions of our thought. They also help us extend our ethical commitments to new, less familiar cases" (Nussbaum 2006, B6).

Ethics of care have much to offer this partial, open-ended, nonnormative attending to flourishment, as ethics of care have emerged from the perspectives often elided by normative ethical frameworks: women, BIPOC (Black, Indigenous, and People of Color), and from those attending to the more-than-human world. An ethics of care is both a value and a practice, and it "builds relations of care and concern and mutual responsiveness to need on both the personal and wider social levels" (Held 2006, 43). Ethics of care is best understood as "a mosaic of insights" (10) rather than a unified, normative theory, which also speaks to its suitability with a relational ontology. However, some core features include: attending to and meeting needs of others for whom we take responsibility; recognizing the value of emotions *and* that care can go wrong (e.g., self-denial, caregiver totalitarianism); questioning abstract rules by emphasizing feminine moral epistemology; and reconceptualizing public and private spheres and reconceptualizing persons as relational—not independent—individuals (Barnes et al. 2015). Theorists of an ethics of care assert that theories of individualism exist because of a (often backgrounded in attention) social network that allows independence. Further, while some ethics-of-care theorists disagree that care should be considered a virtue because virtue can sound individualistic, other theorists have explicitly linked care as a communal virtue and one that is significantly related to *eudaimonia*, or flourishment (discussed more below). Lisa Tessman's (2005) notion of care as a burdened virtue challenges any assumption of pure virtue that is unentangled in the messiness of day-to-day life. Instead of moral purity, Tessman offers an affirmation of *eudaimonic* life in even troubled circumstances (to which I would simply substitute "more-than-human" in place of her use of "human"):

> However, I see the spirit of eudaimonism preserved in a phenomenon that is the opposite of hopelessness and that, perhaps amazingly, survives even the worst sorts of oppression: the affirmation and embrace of life. The choice to go on living, to insist upon life—with its sufferings and its joys—is an existential choice of great significance under oppression, and this choice captures something crucial about eudaimonism. In fact, this phenomenon of affirming life may offer insights into how one is to conceive of *any* human flourishing, for in choosing life one chooses what is at the core of a good life. (168)

Tessman's description is one of particular relevance for the Anthropocene. A collective human response to the Anthropocene that does its best to *live through* the Anthropocene by *living with* and valuing *other ways of life* is perhaps the most radical and *eudaimonic* activity imaginable. However, we still need to look a little closer at what it means to live well with others, which Lori Gruen offers in her notion of entangled empathy. Again, I quote at length Gruen's definition of entangled empathy because it helps tie flourishment to the work of caring in more-than-human communities. It is important to note, however, that Gruen's entangled empathy is already situated in more-than-human ethics of care (Donovan and Adams 2007), so when Gruen speaks of "others," she uses the term as I do to mean not only other humans but more-than-human lives, as well. For Gruen (2015), entangled empathy is:

> a type of caring perception focused on attending to another's experience of wellbeing. An experiential process involving a blend of emotion and cognition in which we recognize we are in relationships with others and are called upon to be responsive and responsible in these relationships by attending to another's needs, interests, desires, vulnerability, hopes, and sensitivities. (3)

Although Aristotle's take on *eudaimonia* excluded the more-than-human world from such flourishment, flourishment need not be anthropocentric. Bringing Gruen and Tessman together, I am describing an ethics of care that addresses the ethical framework's concern with the flourishment of the more-than-human world. In other words, it is a community-care model in which care is considered to be activities, emotions, and cognitions that drive the pursuit of *flourishment* for others. Clearly, I still need to deliver on my earlier promise to elaborate on what *flourishment* is, but before I do, there is one more aspect of the community-care paradigm that needs to be discussed:

who is being cared for in the community-care paradigm (what is the unit of analysis?). Earlier, I wrote that the recognition of distributed agency inherent in relational ontologies places humans in dense more-than-human networks, and these networks all constitute communities that afford different degrees of care for the individuals in those communities. Those last four words are key: individuals in those communities. For me, a community-care paradigm has to be responsive to and accountable for individuals-in-communities.

BASIC UNIT: INDIVIDUALS-IN-COMMUNITIES

When it comes to thinking about how humans should attend to and act in the more-than-human world, there is a long-standing debate between animal activists and environmentalists on whether to emphasize ethical action at the individual level (the individual lives that are often the concern of pro-animal advocates) or at the level of ecosystems (the community wellbeing that environmentalists foreground). Christine Korsgaard (2018) discusses this complication well: It is true that for something to be good, it should be good *for* someone. An individual experiences pleasure and pain, life, and death (18). A species—as a category, not as a collection of individuals making up the species—does not feel. Thus, it is logical to conclude that only individuals have ethical claims on human action.

However, individuals need an environment and a community to flourish. So, it would seem humans need to secure communities and places-of-being for individuals, perhaps shifting our obligation from the individual to the collective. The relational ontology undergirding my articulation of the community-care paradigm points to an indistinction in these spheres of existence. The indistinction of these spheres is a point well made by Thom van Dooren's (2018) articulation of the *dull edge of extinction,* which recognizes that a species could be considered extinct long before the last individual member of the species dies: when that species' unique ways-of-life can no longer be enacted, extinction is already present.

Advocacy campaigns may emphasize individuals or communities, but individuals are always already embedded in communities, and communities are comprised of individual lives. Individuals-in-communities refers to a kind of state in which beings are always already joined together and yet also recognized as individual beings. Further, these communities are of vast spatiotemporal dispersions. As Eduardo Kohn (2013) learned in his ethnographic fieldwork with the Runa community of the Amazon, the self—or individual—is a self in continuity, a continuity that stretches not only across space but also backward and forward in time. Indigenist wisdom articulates that a life *is* a wholeself—heart/emotions, mind/intellect, body/physical action, and spirit/spirituality—in relation to oneself, family (past, present, and future), community, land, environment, and society (Haverkamp 2021).

The notion of individuals-in-communities clearly has ethical dimensions, speaking to the claims individuals make on their communities and communities make of their individuals. Individuals-in-communities also has ecological dimensions, as individuals shape communities and communities shape individuals, and even biological dimensions, as each and every body is a profound example of individuals-in-communities. For example, the human microbiome—the aggregate of all microbiota (i.e., trillions of bacteria, fungi, protists, and other microscopic lifeforms) living on or within human tissue—can be understood as a vast ecosystem that influences a person's immune health and overall wellbeing. At the same time, an individuals' diet and lifestyle influences the diversity and robustness of the microbial community. This example also illuminates that individuals-in-communities is a thoroughly embodied concept.

> A person is not a mind *and* a body. There are not two "things" somehow mysteriously yoked together. What we call a "person" is a certain kind of bodily organism that has a brain operating with its body, a body that is continually interacting with aspects of its environments (material and social) in an every-changing process of experience. . . . In short, "mind" and "body" are merely abstracted aspects of the flow of organism-environment interactions that constitutes what we call experience. (Johnson 2007, 11–12)

With this perspective, individuals-in-communities is fundamentally a concept of naturecultural intra-action. Intra-action is a term, coined by Karen Barad (2007), that acknowledges that individuals are not pre-established but come into being through dynamic exchanges of influence and action. There are no sharp boundaries separating entities, only continuous blendings.

Sometimes, individuals and communities are characterized as separate, distinct realms in order to explore consequences of policies and practices for those realms. When a star tennis player was unable to compete in the Australian Open because the player had not complied with the country's health protocols for the COVID-19 pandemic, Michael Steinberger (2022) of the *New York Times* wrote that at the heart of the debacle was a debate about individuals versus communities. In many cases, it will be important to consider *individuals*-in-communities or individuals-in-*communities*, emphasizing what is at stake for individual lives and communal wellbeing. The point of the hyphenated term, individuals-in-communities, is not to foreclose such discussion, but to remind us that neither term is actually cordoned off from the other. For the sake of understanding the stakes of a policy, it may be useful to discuss how a policy would impact individual autonomy *and* public health *as if* they were distinct spheres of existence.

Choices of individuals impact public health, and public health imposes constraints on individual choices. Even seemingly private health issues do not make sense outside of individuals-in-communities; this is bioethicist Carl Erik Fisher's (2022) point when he writes that it is misleading to think of addiction as a disease. For Fisher, the "addiction as disease" paradigm reduces the complexity of the problem—making it about an individual—and obscures that addressing addiction requires "community support and healing." What both examples bring to the fore is the embeddedness of individuals and communities in a web of relations that reverberates in all directions.

The ethos of individuals-in-communities—that neither term could be (or should be) productively pried apart from the other—is, actually, pretty mundane. At its most basic, individuals-in-communities is the realization of a relational ontology and the recognition that wholes are made of up parts and parts contribute to wholes. Individuals-in-communities helps articulate the power of networks. For example, if each neuron in the brain could store only one unit of information, then even with 86 billion neurons, storage capacity could be a real issue. However, because each neuron can form 1,000 or more connections to other neurons, trillions of connections are possible, and the memory capacity of the brain, while not infinite, is astonishing. This capacity is afforded by neurons-in-networks, or, individuals-in-communities.

Individuals-in-communities is also how people understand teamwork; it is why representation still matters, and it is the ethos evoked in *unus pro omnibus, omnes pro uno* (one for all, all for one). When people discuss the synergy of a good team, it is an acknowledgment of the reverberations occurring with these individuals in this community. The community is more than the sum total of each person's efforts, yet one individual could not be taken away without changing the community. When people discuss the importance of having representation of minoritized populations in films, for example, this speaks to the import of seeing individuals from marginalized communities brought to the fore in privileged communities—like being the hero(es) in a popular culture franchise (Prabasmoro, Budhyono and Muhtadin 2019). Claims for care run in all directions. Individuals are called upon to care for their communities, and individuals make claims on their communities for care.

A final note on the theoretical underpinnings of individuals-in-communities: Individuals-in-communities could also be called individuals-in-relations (Gries and Clary-Lemon 2022, 143), or, in a Deleuzian fashion, selves-in-assemblages. The Deleuzian concept brings to the fore that it is multiplicity and connection all the way down. One need look no further than the microbiome or DNA to be reminded of this. In short, "me" is already "we." When

I say something such as, "I am trying to do a better job of taking care of my health," I am actually referring to lifestyle changes that will impact that entire ecosystem that is my body, as well as the relationships I maintain with other people and, indeed, the more-than-human world. Selves-in-assemblages evokes the plurality and fluidity of these communities nicely, and it avoids some of the loadedness that "individuals" (e.g., "individuals" can conjure Western Enlightenment notions of the autonomous individual) and "communities" (e.g., "communities" often has a positive connotation of mutual interdependence and agency, and this seems to miss the fact that many of the networks we are part of and impacted by do not possess an ethos of mutuality) might be thought to carry.

In the spirit of learning to live well with others, however, I have chosen to work with the term individuals-in-communities because this book emphasizes the community-building capacity of rhetoric and explores how this community-building capacity can help disrupt the kinds of anthropocentrism (i.e., the logics, practices, beliefs, attitudes, and values that separate humans from and articulate humans as superior to the more-than-human world) that fuel an ecologically destructive kind of Anthropocene. This is certainly not to say that *all* anthropologic values, practices, beliefs, attitudes, and values are *bad*, in the Spinozan sense of hampering capacity for life. To dismiss all features of human life in such a sweeping way would just be another form of anthropodenial—to miss out on the many and wondrous capacities and relations of which humans are capable. Movements that call for the eradication of the human species seem to miss that what we see worth preserving in other species is also exhibited by humans, whether it is the familial bonds of elephants, the cleverness of crows, or the creativity of bowerbirds. Rather, a call to attend to the logics, practices, beliefs, attitudes, and values perpetuating anthropocentrism is to suggest that all anthropological practices should be subject to examination: examination of how well they help or hinder humans to live well with others. To conduct such an examination, I first need to offer some sense of what it means to live well with others.

STAYING WITH THE TROUBLE: VULNERABLE FLOURISHINGS

When it comes to the community-care paradigm, I believe the best we can do—as Donna Haraway (2016) suggests—is to be committed to staying with the trouble of (more-than-)human flourishment. And it is troublesome! A commitment to mutual flourishment does *not* mean that all beings flourish equally, at the same time, or under the same conditions. Giraud noted that while nothing may exist outside of relations, "*certain* things might need distance from *certain* relations in order to allow particular realities to be enacted and preserved" (Giraud 2019, 173). The same is true for flourishing; all relations afford and limit certain flourishings for certain beings at certain times in certain contexts. In the United States, for example, the flourishing of kudzu has meant the death of much native vegetation and the life-ways of the fauna that coevolved with the native vegetation. Flourishing is not utopic, and human activity is always impure and partial. Thus, a commitment to mutual flourishment is better conceptualized as an intention (i.e., an ongoing dedication to search for the "hopeful flourishing of *all* beings" [de la Bellacasa 2017, 22] in the struggles of the everyday) rather than a resolution (i.e., a firm determination in which one can *fail* to live up to one's resolve). The ethos of *vulnerable flourishing* is an acknowledgment of what Cora Diamond (2003) refers to as the *difficulties of reality*, including that flourishing is contingent, dependent, and partial.

Although this entire book attends to the matter of vulnerable flourishment—of precarity and wellbeing—I should offer a preliminary accounting of this idea. The idea of *vulnerable flourishment* will be developed throughout this book from a grounded theory approach, this grounded theory is itself developed through engagement with theories of wellbeing and perspectives on vulnerability.

Conceptualizing Wellbeing

One of the first things to note when it comes to (re)thinking wellbeing in the Anthropocene is that that there are multiple theories of wellbeing available, each with different consequences for considering the means and ends of wellbeing. Thom van Dooren (2019) describes (more-than-)human flourishment as the acts of *worlding well,* or the processes of making worlds of more-than-human wellbeing. He describes that what *well* means is not generalizable or final but contextual and dynamic. Still, an overview of wellbeing theorization will productively complicate the concept.

Wellbeing has been pursued as an area of philosophical inquiry, a rhetorical framework, and as a psychological experience of health and happiness. The philosophy of wellbeing addresses the question, "What are we talking about when we talk about wellbeing?" The rhetoric of wellbeing highlights the critical and practical potential of frameworks of wellbeing by analyzing *how and to what extent* discourse and materiality can contribute to wellbeing. The psychology of wellbeing focuses on aspects of resilience and individual pursuits of greater levels of happiness and life-satisfaction.

Within the philosophy of wellbeing, there are theories that are more objectivist (e.g., objective list theories [OLT]) and theories that are more subjectivist (e.g., hedonism), on which I will elaborate shortly. To make matters more complicated, Stephen Campbell (2016) claims that many scholars of wellbeing seem to talk past each other by conflating objectivist and subjectivist concepts of wellbeing. He argues that a person who favors OLT may be considering wellbeing as a *locative* concept, that is, the presence of good things *for* a being whose life can contain good or bad relations and events. In contrast, the subjectivist—such as a proponent of hedonism—may be conceptualizing wellbeing as a *positional* concept, and thinking of wellbeing as being in position to have association with relations that appeal *to* someone. Campbell contends that it would be entirely possible to hold both objectivist and subjectivist positions when the concept of wellbeing is teased out as just described. An example here may help not only to illuminate the locative and positional aspects of wellbeing but also some of the knottiness inherent in the community-care paradigm: "breadcrumbs and wellbeing."

One of my favorite childhood memories is going with my family to a Sunbeam outlet and picking up baked goods—some for my family, some for the ducks at the local park. My parents never balked at buying loaves of bread specifically for the local mallard community, and I still find it touching that part of our food gathering included thinking about and desiring to share with some more-than-human avian neighbors. Though I wanted to, I never got to touch any of the ducks; rather, the joy came from the act of performing care. Both an OLT and hedonist perspective on wellbeing may agree that this child-breadcrumb-duck relation was *good to* (positional concept) many parties involved; there was enjoyment in feeding and being fed. This enjoyment was expansive, as the cashier checking us out enjoyed learning that some of our purchase was destined for the residents of Fairview Park's pond. However, it is also possible for an OLT and hedonists to agree that—certainly for the ducks—the feeding relation was not *good for* (locative concept), in that a diet consisting of low-nutrient foods like bread contributes to metabolic bone disease and other health issues in ducks (The Wildlife Center of Virginia 2022). This example illuminates why the community-care paradigm and a commitment to more-than-human wellbeing requires, as Lori Gruen writes, entangled empathy, both emotion *and* cognition, or the recognition of the entanglement of ethico-onto-epistem-ology.

With this nuance in mind, and thereby troubling any clear and distinct boundaries of the various perspectives on wellbeing, it is still helpful to consider the main schools of thought within the Western philosophic tradition given the substantial archive of wellbeing theorization that has been assembled (and privileged) in this tradition *and* to consider how the question of *the good life* has been taken up by other cultures and outside the discipline of philosophy. The primary schools of thought concerning wellbeing in Western philosophy (in no particular order) are: hedonism, desire-fulfillment theory, OLT, and *eudaimonism*.

Hedonism takes the position that pleasure and pain are the fundamental components of wellbeing, and that given that pleasure and pain are experienced differently by different individuals, wellbeing is ultimately a subjective concept. Hedonism has the benefit of being an explanatory concept of wellbeing that accounts for plurality and multiplicity in a postmodern, post-structuralist era, but it is not without issues, as manifest in—for example—the Nozickian pleasure machine, in which one is hooked into a machine and experiences all and only pleasures.

Most people consider a pleasure machine as insufficient to consider a life well-lived, which points to a shortcoming of hedonism. Desire-fulfillment offers an alternative and considers that what makes a life go well is having our desires satisfied, and what makes a life go poorly is having our desires frustrated. A difference between desire-fulfillment and hedonism is that the fulfillment of desires can include things that are not obviously pleasurable. Desire-fulfillment theory—also subjective—runs into some of the same trouble as hedonism, facing problems like *ill-informed desires* and *malicious desires*.

OLT states that there are things that are good for an individual to have, even if that individual does not desire them; for example, it is good to be an enfranchised citizen even if one does not desire to exercise one's right to vote. OLT has some overlap here with Nussbaum's (2006a) notion—discussed earlier in this chapter—that individuals of all species have a right to the conditions of flourishment for the activities of their species. However, OLT does not offer much to account for subjective experience. Certainly, a person could have rights, privileges, and luxuries and still not experience subjective wellbeing.

Finally, *eudaimonia*, one of the oldest positions on wellbeing in the Western tradition (along with hedonism), contends that the best life is a life well lived, one that develops one's potential. What is important to note in Aristotle's notion of *eudaimonia* is that while it is considered an objective theory—like objective list theories—because Aristotle expounded virtues that contribute to a life well-lived, Aristotle's virtues are also *situated, relational, contextual,* and *lived.* Aristotle claims that we examine good lives not to *know* what goodness is, but so that we may *live well* (Aristotle 2011, 27).

To get a more robust accounting of the good life within a relational and performative ontology and from an ethics of care perspective, it is important to examine good lives beyond the Western philosophic tradition. The *good life* has been taken up in all major religions and is a core component of many cultures. In indigenist wisdom, the good life is a fluid and plural concept, a relational aspiration and practice of living harmoniously with others in the physical, social, emotional, cultural, and spiritual domains. Indigenist wisdom holds that the goal of all human activity should be to create conditions for good lives, and that good lives are only possible in communities (Gudynas 2019; Hidalgo-Capitán, Cubillo-Guevara, and Masabalín-Caisaguano 2020; Van Styvendale et al. 2021).

As mentioned earlier in an endnote, indigenist and Indigenous are terms of massive gathering power that groups many diverse peoples, cultures, and ideologies under an umbrella term that is only legible because of a global history of colonization. Rather than the Indigenous perspective on the good life, it is more appropriate to discuss Indigenous perspectives on the good life.

Ubuntu is a Nguni Bantu term that is often translated as, "I am because we are" (Mwangi 2019, 37). *Ubuntu* directs attention to a communal ethics because "a person is a person through other people" (Mwangi 37). Like *eudaimonia*'s anthropocentric roots, Mwangi contends that while *ubuntu* directs attention to community with other humans, *ubuntu* does not have to be anthropocentric. Understood from a more-than-human web of relations, *ubuntu* is a term that gets at entanglement, interdependency, responsibility, and a commitment to multispecies flourishment.

Buen vivir (good life) is the Spanish rendering and gathering of terms and meanings from many Indigenous communities in South America, including *sumak kaway* (*good life* in Kichwa—a Quechuan language—in the Ecuadorian Upper Amazon), *allin kawsay* (*sweet life* in Quechua language in Peru), *suma qamaña* (*living well together* in Aymara language of the Bolivian Andes), and *kuome mongen* (*a life in harmony with all beings* in Mapuche language of Chile and Argentina) (Hidalgo-Capitán, Cubillo-Guevara, and Masabalín-Caisaguano 2020; Soledad del Villar 2018). *Buen vivir* is a post-development philosophy and politics (Peredo 2019) that articulates as a life well-lived is one understood in a community in a unique place (Bergström 2018; Walsh 2010). Human life exists in kinship and interdependence with the more-than-human world, and both individuals and communities make claims on one another for being treated with respect, responsibility, and compassion (Kemmerer 2017).

Miyo pimâtiswin is the Cree articulation of the good life (of being alive well), a life devoted to balanced and interconnected physical, emotional, spiritual, mental, and ecological wellbeing (Van Styvendale et al. 2021). Wellbeing is concerned with physical health, but health is not privileged as *the* indicator of wellbeing. Rather, health is one part of a wider set of relations necessary for flourishment, and the necessary relations for flourishment include activities like creative expression. The Lakota notion of *wicozani* (wellbeing) holds that a worthwhile life is attained and maintained through awareness of the sacred (i.e., awareness of *wakan* [potentiality, power of life]); through healthy relationships with oneself, family, tribe, and all of creation; through prayer; through the enactment of values of *wacante ognanake* (generosity), *wowaunsila* (compassion), *wowacintanka* (fortitude), *wowauonihan* (honor, respect), *wowahwala* (humility), *woohitike* (courage, bravery), and *woksape* (wisdom); and through the recovery from trauma (Noisy Hawk and Trimble 2019).

Although this brief survey does not do justice to breadth or depth of Indigenous perspectives on the good life, it provides a sketch that—though not fleshed out—makes visible the centrality of a more-than-human community to indigenist wellbeing. What all of these perspectives make clear is that, in addition to posthumanist perspectives, there are also an abundance of alternative-to-humanist perspectives that understand wellbeing as situated in a more-than-human world. Wellbeing is not a human affair but an Earthly, or even cosmic, affair.

The final aspect of wellbeing from a community-care paradigm that needs to be addressed comes from rhetorical contributions to wellbeing theory (and I will be turning to the worldbuilding features of rhetoric more fully in the next chapter). The relationship between rhetoric and wellbeing has been debated since the ancient Greeks theorized the nature, scope, and function of rhetoric. For Plato, rhetoric was a sham art that hindered one's ability to live a good life (Plato 2001, 97–98); for Aristotle, rhetoric is *a means* of achieving the good life, asserting that "all forms of exhortation and discussion are concerned with this [*eudaimonia*] and with the things that contribute, or are opposed, to it" (Aristotle 1991, 57). More recently, Gallagher and colleagues (2011) extended the rhetoric of wellbeing by attending to nonlinguistic, visual aspects of wellbeing and articulating how visual wellbeing can function as a critical framework for analyzing and assessing artifacts and objects. Crucially, while most theories of wellbeing attend to wellbeing from the perspective of an entire life (i.e., a recounting of sorts of an entire life lived), Gallagher and colleagues explored how episodes of sensorial wellbeing might contribute to cultivating more resources for long-term wellbeing. The episodic possibilities for *eudaimonia* are foundational to the notion of vulnerable flourishment. Gallagher and colleagues provide a way of thinking about *eudaimonia* as embodied, as experiences of flourishing. What remains, then, is to consider how and to what extent flourishment is useful for thinking about the wellbeing of the more-than-human world.

Nussbaum (2006a) asserts that attending to more-than-human wellbeing is an act of justice (392); I would add that it is also an act of care. Her approach to more-than-human flourishment—the capabilities approach—recognizes a wide range of types of dignity (e.g., sensation, affiliation) and corresponding needs for flourishing (e.g., access to sense data, opportunities to form attachments). The core ethical intuition of the capabilities approach is that the more-than-human world is entitled to "a life worthy of the dignity of each creature" (353).

Recall that for Nussbaum, nonhuman animals are entitled to lives of flourishment, lives in which they can live in dignity as the kinds of creatures they are. Flourishing, then, has both evaluative and ethical dimensions. The ethical dimension of more-than-human flourishing does entail human interventions (and, of course, humans are *always already* intervening in the more-than-human world), but Nussbaum resists offering a homogeneous, static intervention that would enable more-than-human flourishment. Nussbaum argues for ensuring adequate conditions in which others can flourish, which means we have a positive (not only a negative[6]) duty to nonhuman animals. Nussbaum also stresses that flourishing is plural and that plans must always be considered partial and subject to revision. My point here is not to provide an assessment of the capabilities approach to flourishment;[7] rather, I am interested in how this line of thought (i.e., that more-than-human flourishment is a legitimate ethical concern for humans) directs attention to matters of more-than-human flourishment and what it would mean, in the words of Matthew Calarco, to exist in "a lively, flourishing, multispecies zoo-polis" (Calarco 2021, 11)

In this book, I will be returning to wellbeing and how disclosure, transformation, and infrastructuration help elucidate flourishment. For now, consider that within the community-care paradigm, wellbeing is taken to be a more-than-human affair, in which the wellbeing of each being is intrinsically valued, and human wellbeing is understood to be indistinct or inseparable from the material world in which humans reside. Wellbeing is taken to be situated and interdependent, and is better understood as possessing attributes of both objective and subjective perspectives rather than as a bi-polar model ranging from objective theories on one end and subjective theories on the other. Additionally, wellbeing has both ethical and evaluative dimensions and, finally, is partial and dynamic with varying degrees of durability. Flourishment, of any kind, is always vulnerable.

The Vulnerability of Flourishment and the Flourishment of Vulnerability

Vulnerable flourishment takes these two terms together, as interlocking, entangled—tying flourishment to the existence of vulnerability and vulnerability to the possibility of flourishment. This may not sound ideal. "Durable flourishment" or "perpetual flourishment" might seem a more optimal goal. However, flourishment is neither given nor guaranteed; it must be continuously cultivated. In circumstances of oppression and crisis, virtue and *eudaimonia* show up differently. Choosing life—human and more-than-human—and endeavoring *to world well* in a time of precarity are, as Tessman (2005) would say, at the core of a good life (168).

In alignment with Tessman, I see vulnerability as the condition in which any and all flourishing occurs and resist seeing vulnerability as a pessimistic or nihilistic state of affairs. Vulnerability is inherent in the act of choosing life, which is not to say that vulnerability is equally distributed among all beings or even among all humans; oppressive, hierarchical systems make certain subject-positions more or less vulnerable to a number of injustices ranging from housing and food insecurity to health disparities. Thus, vulnerabilities differ, and while vulnerability may be the basic condition of embodied life and the condition in which flourishment may occur, the variations in kinds and degrees of vulnerabilities mean that not every individual gets to experience wellbeing. Vulnerability may be a biological norm but it is also a sociopolitical matter of injustice.

Vulnerability is also part of the flourishment of relationships; in being vulnerable with another, people open themselves not only to the possibility of pain but also to the possibility of radical and reparative acceptance (Rosenfeld 1979). This is what social worker and author Brené Brown (2012) refers to as the power of vulnerability; Brown argues that there is neither opportunity nor intimacy without vulnerability. "Vulnerability is the birthplace of love, belonging, joy, courage, empathy, and creativity. It is the source of hope, empathy, accountability, and authenticity" (34). As I show in chapter 4 when I discuss disclosure and how people reveal themselves—including dietary practices—to one another, vulnerability is also a condition of flourishment within human relationships. Here, the implication of flourishment in vulnerability is already made by Brown, but I want to continue on this path for a moment to (1) point out that vulnerability is not only an act of intention (i.e., choosing to be vulnerable) but the very condition of embodiment, and (2) make explicit the more-than-human aspects of vulnerability and flourishment.

For scholars as diverse in their thinking at Martha Nussbaum, Cora Diamond, Donna Haraway, and Val Plumwood, vulnerability is key in recognizing nonhuman animals as fellow creatures (Wolfe 2010, 62). It is in the mortality and finitude of embodiment that humans can tap into the reality of their animal bodies. A recognition of shared embodiment and vulnerability can facilitate empathy with the more-than-human world, and this recognition is not only cognitive but also embodied. When Plumwood (2012) was attacked by a crocodile, her profound experience of embodiment and vulnerability led to a greater appreciation of her effort to live, an effort shared by the crocodile. She and the crocodile were both—in their own unique vulnerabilities as the hungry predator needing nourishment and the potential prey needing to avoid predation—choosing live, to flourish as the kinds of creatures they were.

I do not believe we need to share in an experience as harrowing as Plumwood's to recognize the shared mortality and finitude of embodiment. The existential threats of the Anthropocene provide "shared conditions, overlapping experiences, and common fates" (Calarco 2020, 31) conducive to recognizing mutual vulnerability. Beyond that recognition lay possibilities for commitment to mutual flourishment, that is the "genuine flourishing" that is to be found in living well with others (Calarco 2020, 33). This book is dedicated to surveying places and practices where rhetorics of vulnerable flourishment take shape so as to provide some ideas and techniques for *worlding* well (van Dooren 2019, 9).

STORY TIME: EEEEERRRPPP, EEEERRPP, EEEERRPP, EEEERP

For about a week in late 2016, I kept my companion chickens confined to the area of their coop, which included a small, enclosed run. We had just experienced the second hawk predation on our chickens, and I did not know what else to do. After the first hawk attack, my partner and I studiously researched all the strategies for preventing a hawk attack. We had hung up metal pie pans to create glares that would discourage a swoop attack, put out owl statues to indicate the presence of another raptor, and even left out a small radio playing NPR so the hawks might choose to avoid the human sounds. When the second attack occurred, we were out of ideas for how to protect these chickens but still felt it was our job to protect them. So, we decided to keep them in their coop and run.

It's not that bad, we told ourselves. They are safe. They can still scratch at the ground and get in a dust bath. They still had more run space than any chicken in a confined feeding operation, and we made sure to add lots of food and items, such as a xylophone, for enrichment. This was surely the best compromise between *letting chickens be chickens* and *protecting chickens from danger*.

During the week of confinement, the hens made long, persistent calls of "eeeeerrrppp, eeeerrpp, eeeerrpp, eeeerp" as they passed back and forth in front of the door and along the wire sides separating them from the rest of the world. The first day, I told myself that they would stop as the day went on, but they only stopped when they put themselves to bed for the evening. That night, I told myself that it was a "first day" reaction, and surely tomorrow, they would be adjusted to the new arrangement. Nevertheless, they persisted.

I found myself avoiding the chickens and even the outdoors on days three and four. The incessant pacing and calling—behaviors I never saw them engage in when free-ranging—was distressing. I told myself I was being anthropomorphic, projecting my own dissatisfaction with the arrangement onto the chickens and interpreting the pacing and unique vocalizations through that lens. Also, so what if the chickens wanted out? What do the chickens know about their safety? I was the one aware of all the dangers they faced; I needed to step in and protect them from their own uninformed desires to free-range. I had a whole storehouse of tropes I could use to justify my enforced confinement.

Taking fresh food and water to the chickens on the fifth morning, my storehouse was vacant. The tropes had reached their expiration date and were no longer useful as justifications. I looked at and listened to these chickens, and I responded, "But *you are not safe* out here!" Their behavior did not change. My eyes swelled with tears. I open the door and increase their vulnerability. I keep the door closed and decrease their flourishment. If I kept them cooped, through learned helplessness they might eventually give up their pacing and calls. But I would know.

John Shedd (1928) wrote that a ship in a harbor is safe, but, in important ways, it is not a ship. To flourish *as a ship*, it has to go to sea. I knew this line for years, but I did not *feel* this statement until my companion chickens called out, day after day, "Eeeeerrrppp, eeeerrpp, eeeerrpp, eeeerp!"

As I opened the door to a forty-five-degree angle, chickens began pouring out of the coop. Some chickens ran toward the goat shed to take dust baths in their most frequented location. Some chickens began scratching and pecking at the ground surrounding the large oaks near the coop. Some chickens just seemed to be taking a walk. Each was being a chicken-in-the-world with all its delights and dangers.

NOTES

1. In *Philosophy of Biology*, Peter Godfrey-Smith (2014) troubles any easy definition of what makes a species, including the phenetic view of *overall similarity* and Ernst Mayr's *biological species concept* (a concept that defines species as members of an actual or potential reproductive community).

2. The "Indigenous perspective"—a misnomer because it conflates all of the numerous Indigenous communities around the world into one category that is contrasted against "the Western perspective" (which is also more diverse than often acknowledged, including by me)—offers an abundance of community-care practices, prayers, and phrases. It is beyond this book to offer a robust accounting of the multitudinous community-care aspects of the world's many Indigenous communities, but here are two places one can look to get started: Berkes (1993) and LaDuke (2020).

3. Of course, *Indigenous* is a problematic collective noun that gathers disparate peoples from all over the globe—each with place-based values, cultures, and epistemes—under the negative umbrella of "not-settler."

4. I am recounting this story for the purpose of illustrating multiperspectival knowing; however, I should also note that this story includes the human then trapping the jaguar in the house. This is not a simple story of more-than-human cooperative activity.

5. The problem with "knowing-that" and the authorization to act on others undergirds much intra-human oppression, as well. There are many excellent book-length treatments that speak to these issues, but a few select books that speak to how "knowing-that" can permit human and more-than-human atrocities include: Boisseron 2018, Kim 2015, Deckha 2018, Mwangi 2019, and Taylor 2017.

6. A negative duty refers to a duty *not* to harm or injure another.

7. Scholars have critiqued the capabilities approach as (1) furthering an ableist understanding of wellbeing, in which flourishment is associated with physical abilities, such as capacity for movement, and (2) being too humanistic to affect meaningful change in anthropocentric logics. I want to keep these critiques in mind as discussing hazards in the capabilities approach (and, of course, all concepts, frameworks, and theories have hazards), while retaining Nussbaum's impulse that flourishment is a more-than-human affair.

Chapter 3

Worldbuilding Digital Rhetorics

In the beginning of *Slaughterhouse-Five*, the narrator recounts a time when he shared his desire to write a book about Dresden, Germany, in World War II. The interlocutor, movie-maker Harrison Starr, inquires if the narrator intends this to be an anti-war book. The narrator responds, "I guess" (Vonnegut 2019, 5). Starr replies, "'Why don't you write an anti-glacier book instead?'" (5). Lest there be any confusion of Starr's meaning, the narrator clarifies for the readers, "What he meant, of course, was that there would always be wars, that they were as easy to stop as glaciers" (5).

Decades after *Slaughterhouse-Five*, an abundance of books—as well as scientific and environmental humanities scholarship on climate change—call the inevitability of glaciers into question. A surfeit of popular science books have detailed the exigences of the Anthropocene. Displayed on local bookshelves or under the "Customers Who Bought This Book Also Bought" section of a website are titles like: *The Uninhabitable Earth: Life After Warming* (Wallace-Wells 2019); *This Changes Everything: Capitalism Versus Climate* (Klein 2014); *Losing Earth: A Recent History* (Rich 2019); *Under a White Sky: The Nature of the Future* (Kolbert 2021); *The Great Derangement: Climate Change and the Unthinkable* (Ghosh 2016); *The New Climate War: The Fight to Take Back Our Planet* (Mann 2021); *How to Prepare for Climate Change: A Practical Guide to Surviving the Chaos* (Pogue 2021); *This Is the Way the World Ends: How Droughts and Die-Offs, Heat Waves and Hurricanes Are Converging on America* (Nisbet 2019); and *Learning to Die in the Anthropocene* (Scranton 2015).

In the Pulitzer Prize–winning *The Sixth Extinction: An Unnatural History*, Elizabeth Kolbert suggests that humanity's most lasting legacy might be our responsibility for a mass extinction of biodiversity and the only mass extinction event catalyzed by the ways-of-life of a single species. "Right now, in the amazing moment that to us counts as the present, we are deciding, without quite meaning to, which evolutionary pathways will remain open and which will be forever closed" (Kolbert 2014, 268). In a rich visual metaphor of knitting, Emma Marris (2021) directs attention to the millions of years of interwoven lives, processes, and landscapes that gave rise to our present state of biodiversity. If the timescale of weaving diverse forms and ways of life is so long and laborious, it would seem reasonable "that *unknitting* would take time too. But no. Death is easy, like slipping into the bath" (Marris 2021, 251).

At risk in this great loss of biodiversity is not just the charismatic megafauna that animate many conservation campaigns but also common species—including humans (MacKinnon 2018). As Pezzullo and Onís (2018) put it, the "vulnerability of Earth's capacity to sustain our species poses an exigence we all would do well to hear and act upon" (103). In short, the ecological effects of the Anthropocene are *bad*, bad in the Spinozan sense that they decrease the power of living for humans and other beings (Deleuze 1988).

The sobering titles above speak to these devastating ecological effects and announce a dramatic exigence the Anthropocene poses: an urgency to change sociopolitical and economic practices that are incommensurate with maintaining a healthy, diverse biosphere. Yet, like Mr. Starr's response to the idea of an anti-war book, the titles also seem to announce something of the inevitable (even if within the pages of these books, many authors hold hope for intervention). The Anthropocene and climate change are understood in stark adjectives and verbs: uninhabitable, loss, derangement, war, chaos, world-ending, preparing to die. Perhaps, if Vonnegut were writing today, the narrator would say he was writing an anti-war book, and Mr. Starr would reply, "Why don't you write a pro-glacier book instead?" "What he meant, of course," the narrator would then comment, "was that there would always be wars, that they were as easy to stop as the melting of glaciers."

Returning to that imagined physical or virtual book shelf containing climate change books, a potential reader might take note of some other titles, too: *No One Is Too Small to Make a Difference* (Thunberg 2019); *The Future We Choose: Surviving the Climate Crisis* (Figueres and Rivett-Carnac 2020); *We Are the Weather: Saving the Planet Begins at Breakfast* (Foer 2019); and, more humbly perhaps, *Living in the Anthropocene: Earth in the Age of Humans* (Kress and Stine 2017) and *Urgency in the Anthropocene* (Lynch and Veland 2018). If all of the titles from both lists could be grouped together under some umbrella like "Anthropocene and Climate Change," then these last five titles introduce a different take with words like "difference," "choose," "saving," "living," and even "urgency." These titles introduce the Anthropocene and climate change as *contingent*, and that, Aristotle tells us, makes the Anthropocene and climate change a rhetorical matter.[1]

Thom van Dooren (2018) argues that we should not think about "species survival" merely as preserving some individuals of a species. Species survival is also about the preserving of diverse ways-of-life, human and more-than-human. Both climate change and its remedies pose crises for the surviving and thriving of the more-than-human world. To attend to matters of survival and flourishment, it matters whose rhetoric we attend to and what actions those rhetorics encourage; in other words, it matters what rhetorics help the Anthropocene to take shape and what rhetorics help to shorten and thin the Anthropocene (Clary-Lemon 2019, 22).

The nature (what kind of thing rhetoric is), scope (what entities and processes can be considered rhetoric), and function (what ends rhetoric achieves) of rhetoric have been debated for millennia. I will in no way attempt to define or stabilize the nature, scope, and function of rhetoric here, as that would go against both my ethico-onto-epistem-ological position of dynamic relationality and against what I believe is key to rhetoric's longevity and importance as an area of study: the worldbuilding capacities of rhetoric. Instead, in this chapter I discuss some rhetorical concepts that are foundational to the worldbuilding capacities of rhetoric and conclude by addressing the importance of surveying the worlds being built in digital spaces and with digital technologies.

WORLDBUILDING RHETORICS

Not just any rhetoric will do to meet the biosphere-preserving demands of the Anthropocene. A rhetoric founded on conceptions of predetermined, agential rhetors exclusively *speaking* or *writing* to a passive audience—in which success and failure is based on the realization of the rhetor's intent—is insufficient. Not only have such static and binary notions of agency, authorship, and audience been thoroughly critiqued and found wanting in the works of Roland Barthes, Jacques Derrida, and Michel Foucault, but the *means* of rhetoric in this conceptualization are limited and understated. The problems of the Anthropocene necessitate a rhetoric that is more complex than a model of persuasive intent and require a model that recognizes a plurality of means and rhetorical agents, and more subtly, recognizes nuance in rhetorical effects that exceed authorial persuasion.

The expansion of rhetoric's terrain from public address and composition to include visual rhetoric, material rhetoric, performance, and space-and-place provides a strong foundation for a complex, more-than-human rhetoric. More-than-human rhetorical studies have looked at nonhuman animal rhetorics and rhetoricians (Greenwalt 2018; Gruber 2018; Kennedy 1992, 1998; Parrish 2014), the rhetorics of material practices (Dickinson 2011; Herndl et al. 2018; Senda-Cook and McHendry 2018), media and technology rhetorics (Arola 2017, 2018; Rickert 2013; Springer and Goggin 2013), governing apparatuses (Greene 1998), land rhetorics (Arola 2018; Carpenter 2013;

Kirsch 2013), matter's intrinsic rhetoricity (Opperman 2014), and rhetorics of vibrant environments (Davis 2011, 2014; Rickert 2013). In recent decades, scholars have expanded their study of the means of persuasion and identification to include coffee shops (Dickinson 2002), quilts (Blair and Michel 2007), memorials (Balthrop et al. 2010), museums (Dickinson, Ott, and Eric 2005), parks (Zagacki and Gallagher 2009), grocery stores (Dickinson and Maugh, 2004), viral images (Gries 2015), activist image events (DeLuca 1999), local food events (Senda-Cook and McHendry 2018), maps (Prelli 2006), and urban markings (Senda-Cook et al. 2018). Greater attention to rhetorical agents has diversified the study of suasory means and made *persuasion* a more nimble term. For example, while some urban markings are human-constructed, humans are far from being the only rhetors engaged in place-making: there are the coyotes of Chicago who lay their own claims to territory, resist human designs and intentions for how to move through a city, and disrupt understandings of what a sandwich shop is (Seegert 2014). Further, humans are not the only species who create elaborate, multimedia displays that merge the "natural" (e.g., flower petals, berries) and the "technological" (e.g., toy cars, wrappers) (Rosenfeld 2021).

The expansion of the study of rhetoric to these different terrains sheds new light on these phenomena and offers insight into how rhetoric functions to shape species (Kennedy 1992; Parrish 2014; Wheeler 2014), shape cultures (Herndl et al. 2018; Opperman 2014; Senda-Cook and McHendry 2018), shape selves (Kennedy 1992), shape knowledge (Bogost 2007; Springer and Goggin 2013), shape relationships (Davis 2011; Greenwalt 2018; Seegert 2016)—in short, to shape worlds. As rhetorical critics and theorists recognize rhetoric's embeddedness in the more-than-human world, their scholarship draws attention to rhetoric's embeddedness in and co-constitution of ecologies of spaces and places, cultures, politics, media, and embodiment, including affect and emotions, physical vulnerability, sensorium, and biological processes. This book joins a host of other projects that both attend to the ways rhetoric is of, in, and making of world(s) (Barnett 2019; Clary-Lemon 2019; Rickert 2013).

One of the clearest examples of rhetoric's worldbuilding capacity—an example which will help to clarify the term "worldbuilding" as I am using it—comes from Plato's (2001) famous account in *Gorgias* of the debate on rhetoric's nature, scope, and function between Socrates and several sophists, including Gorgias. Early in the pseudo-dialog (*pseudo* because Plato provides some lengthy monologues for his mentor, Socrates) between Socrates and Gorgias, Gorgias proclaims that the full power of rhetoric can be seen in the bridges of Athens. These bridges did not appear on their own, nor did engineers and carpenters randomly start construction somewhere. The infrastructure of a city, which shapes a city's and people's movements, is a matter of deliberation. Infrastructure is something of a key term or threshold concept for unpacking the Anthropocene. The infrastructures that undergird the activities and consequences of the Anthropocene—from industrial agriculture to global supply chains to food deserts to nuclear waste facilities—are not just passive material and discursive arrangements but sites that are constructive of ways-of-living (the bridges of Athens made possible new routes and ways of travel, new ways-of-living, even structuring time). As such, all peoples and nonhuman beings, present and future, have legitimate ethical claims on the shape these infrastructures take and the worldbuilding rhetorics—including the ideologies, affects, politics, and economics—they mobilize.

Van Dooren (2019) writes that ethics is the work of "crafting flourishing worlds, of *worlding well*" (9). This succinct statement captures well the entwinement of rhetoric and ethics. The art of crafting flourishing worlds—that is, of both mobilizing and partially stabilizing passions, thoughts, and actions—are rhetorical processes. Worldbuilding does not have to be as obviously world*building* as in the case of bridge building and physical constructions that change the shape of land. Worldbuilding also refers to, for example, works of science fiction (Seraphin 2020; Vint 2010) that allow people to imagine other worlds and ways of being, as well as the activities of community-organizing (Imarisha 2015, 3) that allow people to imagine different possibilities in their own communities. What is important across these examples is their rhetorical nature, their capacity to act on and with bodies to make certain possibilities for action and relation more likely and others less likely.

When I speak of the worldbuilding capacity of rhetoric, I am not speaking of a compulsion; the worldbuilding capacity of rhetoric often works slowly and subtly. For example, Donnie Sackey (2022) describes the *invention* of bedbugs through laws, capitalistic practices, and class and racial discrimination. The complex arrangement of tenancy laws, industries dedicated to pest control, and rhetorics of sanitation have (1) rendered the presence of members of the Cimicidae family, to which bedbugs belong, to be a sign of uncleanliness, often with a moral valence, and (2) stabilized members of the Cimicidae family as the mass, killable category of "pest." What Sackey's work demonstrates is the *invention* of bedbugs as they are understood and encountered today, or, how the world of *bedbugs* came to be.

There are a number of rhetorical concepts that reveal rhetoric's worldbuilding potentialities. For now, I am naming these concepts and providing a brief description of each in Table 3.1; shortly, I will explain these concepts more fully in a generative example; that is, I will show how these concepts work together to explicate the worldbuilding dimensions of a critically made trashcan. These concepts include (but are in no way limited to) the constitutive nature of rhetoric, the material recalcitrance of rhetoric, rhetoric's sensorium,

Table 3.1. Worldbuilding Rhetoric Concepts

Concept	*Worldbuilding Potentialities*	*Theorists*
The constitutive function of rhetoric	The ability of rhetorical entities—whether linguistic, visual, or material—to create character and community or to bring about a collective identity	Charland 1987
Material rhetoric	The suasory functions of matter's spatial organization, mobility, mass, utility, orality, and tactility	Blair 1999
Rhetoric's sensorium	The suasory force of perceptual, sensual encounters in/of the world	Hawhee 2015
Affective and asignifying rhetoric	The social mobilization of preconscious feelings through the forces, effects, and actions of rhetoric	Ingraham 2020; Muckelbauer 2021
Procedural rhetoric	Claims on the world made through embodied processes and constructed models	Bogost 2010
Embodiment and embeddedness of rhetoric in more-than-networks of relations	Rhetoric dispersed across pluralist networks of relations and things, each with gradations of agency of their own	Rickert 2013; Seegert 2016
Energy of rhetoric	The emotional, physical, coded, and decoded aspects of rhetoric	Kennedy 1992; Ingraham 2018

Table created by author.

the affective and asignifying aspects of rhetoric, the suasory impact of procedures, the embodiment and embeddedness of rhetoric in more-than-human networks of relations, and the energy of rhetoric.

In discussing these concepts together, I do not mean to imply that they share a fundamental understanding of the *nature* of rhetoric. For example, Parrish (2014) describes rhetoric as a more-than-human communicative act that involves the use of intention to achieve some adaptive end, while Rickert (2013) describes intention as only one aspect of some forms of rhetoric. However, both Parrish's and Rickert's works—and all of the concepts that I will briefly trace—underscore how rhetoric mobilizes certain behaviors, relations, and/or ways of being in the world. Both Parrish's and Rickert's works express how suasion participates in worldbuilding.

Rhetoric's constitutive capabilities speak directly to the worldbuilding capacity of rhetoric. The constitutive capability of rhetoric refers to the ability of rhetorical entities—whether linguistic, visual, or material—to create character and community or bring about a collective identity. As Maurice Charland (1987) describes it, "the very existence of social subjects (who would become audience members) is already a rhetorical effect" (133). Although Benedict Anderson never used the term "constitutive rhetoric" in his description of *imagined communities* (i.e., a nation as a socially constructed community that is realized by individuals who perceive themselves as part of a whole), Anderson's depiction of the way written language, maps, and the census all function as tools for creating social subjects who would become a mass audience is an excellent example of the constitutive—and worldbuilding—*dunamis* (i.e., Greek for potential, ability, and power) of rhetoric.

If language were the only or primary way that rhetoric constituted subjects and communities, that would be challenging enough to analyze. However, the constitutive *dunamis* of rhetoric exceeds (human) language. Indeed, the constitutive *dunamis* of rhetoric is more-than-human. Orcas have dialects (Deecke et al. 1999), and whales have been found to change songs through cultural exchange with other whales (Helweg et al. 2005). To illustrate the more-than-human *dunamis* of rhetoric and how materiality, sensoriality, affectivity, asignification, procedurality, embodiment and embeddedness, and energy articulate together to constitute a particular character or community, I offer a brief analysis of my foray into making an alternative trashcan I named *Talking Trash.* The analysis should help make clear how these concepts co-participate in worldbuilding.

WASTE WORLDBUILDING

Talking Trash is an attempt to interrogate and intervene in human waste-world-making. Waste-world-making is important to query because these practices are a hallmark of the Anthropocene's materialization as a new layer of the detritus of industrial human activities. Talking Trash is a small receptacle I made by weaving a basket from strips of environmental articles from magazines I received in the mail by virtue of donations or memberships to various animal and environmental advocacy groups. Rather than a comprehensive review of literature of the trends and themes present in these articles, my aim was to create a different kind of remixing of them: the material existence of these articles would be interlocked and used to craft a receptacle that would encourage different trash practices. After creating the basket, I set up a Twitter account (@Talking_Trash_) to document what I discarded in the basket.

The materiality of Talking Trash participates in building a world in which trash aesthetics, or the sensorial encounters of waste, are foregrounded. There is no lid to hide the comingling and decomposing of the can's contents, no barrier to prevent an odor from binding to the receptors in nearby noses. Not only is Talking Trash uncovered, it also uses a weaving technique that creates open spaces throughout the receptable. While opaque containers and dense material barriers characterize most domestic, commercial, and industrial waste receptacles, *Talking Trash* leaves little to the imagination—or the sensorium—regarding the materiality of what is being discarded. The porosity of Talking Trash offers a glimpse into what happens at the landfill, where organics combine with rainwater and, in mixing with the other trash, create the toxic liquid known as leachate.

As material rhetoric, Talking Trash rewards attention and energy. As an object, it takes energy to engage with Talking Trash as a waste receptable because it jams the neural network of predictive simulation. Predictive simulation refers to the theory of brain functioning that describes the brain as continually working to create probabilistic predictions based on mental models that are constantly being generated and updated by comparing predictive sensory input to actual sensory input. For many, a robust predictive simulation of waste practices has been built through prolonged engagement with mainstream waste-world-making, and Talking Trash violates the neural algorithm's expectations. Indeed, it takes energy to even decode the paper-woven basket as a potential site of waste storage: it requires a process of reorienting or reattuning—not unlike when one encounters an ambiguous image (or reversible figure) and must re-orient to see the other distinct image form(s) that one did not initially see. In a network of relations with other trashcans and materials and practices, Talking Trash attunes people to trash differently.

The Twitter-based aspect of Talking Trash, @Talking_Trash_, is not a digital representation of the material Talking Trash receptacle. Rather, it is an intervention into the privacy of waste-world-making. Instead of keeping trash private, @Talking_Trash_ makes trash and trash practices public and makes visible a wider web of waste relations (i.e., the sociocultural matrix in which the alternative trashcan exists). @Talking_Trash_ has 2 followers, unlike @WasteManagement, the Twitter account associated with the largest waste management company in the United States, which has 47,000 followers. @Talking_Trash_ highlights the ambiguity of the *n* in the one-to-*n* communication afforded by social media: the attempt to publicize does not guarantee a receiving public(s).

Although there is a substantial difference in the digital network size of @Talking_Trash_ and @WasteManagement, the 47,000 followers are only a small fraction of the over 20 million people who engage Waste Management, Inc. services residentially, industrially, municipally, and commercially.

The header photo on @WasteManagement visualizes well the disjuncture between use (of the service) and engagement (with the multi-billion dollar industry and the "what happens next" of the refuse). In the image, there is a row of large, opaque containers as one might see lined along a suburban street. One green container is being lifted, but is *not* seen being dumped, toward a pristine truck sporting an ever larger, opaque bin. The procedures of daily trash practices—depicted in this photograph—obscure their embeddedness in the material world in which trash never is truly "away" from, only elsewhere (Leonard 2010). The publicity of trash practices is not visible in the public-facing social media account of Waste Management, Inc. Even if the practices were highly visible, the social media account is not highly accessed. Waste Management, Inc. qualifies as a social media "microinfluencer" (i.e., having more than 1,000 but fewer than 100,000 followers), which makes for an interesting contradiction between online/offline influence. For the twenty million people being serviced by Waste Management, Inc., the company is a significant influence on day-to-day life. A few missed days of service would quickly acquaint people with the volume of their refuse.

On Instagram, artist and activist Benjamin Von Wong has over one hundred thousand followers, what some consider "mid-tier" influencer status. Von Wong combines fantasy and photography to produce provocative images, such as "This ocean is made of 10,000 plastic bottles," which features mermaids swimming in a sea of plastic and "#TurnOffThePlasticTap," which shows a large faucet flooding an industrial building with plastic while a small group attempt to capture the plastic with a woefully inadequate recycling bin. While the tweets associated with @Talking_Trash_ get at trash practices through linguistic means (e.g., providing information like trash statistics and suggesting alternatives like composting), Von Wong's images work in more asignifying and affective registers—exerting a compelling force, evoking particular responses.

Talking Trash, @Talking_Trash_, @WasteManagement, and "#TurnOffThePlasticTap" make clear how all waste receptacles—or lack thereof—participate in the *inventio,* or invention, of certain kinds of activities and relations—certain kinds of worlds. The header image of @WasteManagement discloses a well-ordered and contained world of refuse. TalkingTrash and @Talking_Trash_ still work to "contain" trash, albeit in a more porous manner, even as it removes any barriers from encountering the sights and smells of what it is being discarded. Neither @Talking_Trash_ nor @WasteManagement speak to the scale of the problem or help audiences visualize how a geologic layer of detritus is one of the markers of the Anthropocene. In contrast, "#TurnOffThePlasticTap" envisions a chaotic world in which plastic refuse will soon overfill containment capacity, drowning people and industry in its wake. If Talking Trash offers a glimpse of the toxic mixture that occurs at the *landfill*, "#TurnOffThePlasticTap" manifests the toxicity of a *land full*.

Despite their differences, both "#TurnOffThePlasticTap" and @Talking_Trash_ gesture toward the sociality of waste-world-making through their uses of the hashtag and at symbols. They speak into a world of hashtag activism (de Kosnik and Feldman 2019; Jackson, Bailey, and Welles 2020), in which social media contribute to civic engagement and social movements. Both the @ and # symbols serve to join in conversations and build networks. As agents in disclosing trash differently, both "#TurnOffThePlasticTap" and @Talking_Trash_ help to create what Chris Ingraham (2020) calls an affective commonwealth, or "a shared sense of what it feels like to be alive at the present time, but as if that feeling were a resource anyone could draw on to make sense of their worlds and to affirm more sustainable ways of being interconnected within them" (5). Both "#TurnOffThePlasticTap" and @Talking_Trash_ are engaged in worldbuilding activities that share a sense of urgency to act in light of the environmental and existential crises of the Anthropocene.

The discussion of worldbuilding rhetorics offered through this examination of Talking Trash and @Talking_Trash_ is highly resonant with Jennifer Clary-Lemon's articulation of *new materialist environmental rhetoric*. In *Planting the Anthropocene*, Clary-Lemon (2019) proposes a new materialist environmental rhetoric that blends ambient rhetoric, materialism and material ecocriticism, and critical affect studies, and shares a relational ontology with postcolonial and Indigenous paradigms. *New materialist environmental rhetoric* (NMER) directs attention "to the action of rhetorical forces, effects, movement, and relations on a multitude of bodies" (64). For Clary-Lemon, such a rhetoric includes all our relations—human and more-than-human—and attends to the "sites, words, things, processes, and sensoria that invent the Anthropocene" (5). My focus on worldbuilding rhetorics aligns with this approach: I seek rhetorics that build compatible worlds with Clary-Lemon's. The distinction between worldbuilding rhetorics and NMER, then, is that NMER is a kind of rhetoric (i.e., environmental rhetoric) informed by a particular (though expansive) paradigm (i.e., new materialism), and worldbuilding rhetorics refers to the capacity of concepts and processes to (de)stabilize ways-of-life.

As mentioned earlier, this book is part of a larger body of work that attend to rhetoric's worlding—how rhetoric shapes world(s), how world(s) shape rhetoric—with the shared aim of coexisting well with earthly relations. Given the emphasis on earthly relations, it may seem paradoxical that I focus on digital rhetorics in this book. I will attend to this paradox now.

WORLDBUILDING IN DIGITAL SPACES

The three rhetorics I examine closely in this book—the invitational rhetoric of snake identification boards, the disruption introduced by vernacular plant-based visual rhetorics, and the (de)colonial projects that (un)ravel the relations of the Anthropocene—are all digital artifacts that re-present and articulate different more-than-human relations on social media. For a book invested in politics and ethics of living in/with the more-than-human world, digital spaces may not be the most obvious of spaces to look. However, Michele Kennerly and Damien Smith Pfister (2018) make a compelling case that digital rhetorics have antecedent, analogic, heuristic, conventional, and renewal relationships with other rhetorics. Further, as anthropologists have pointed out, fieldsites are fragmented and exist across both digital and physical spaces, and digital ethnography offers a valuable, legitimate, and unique way of "being there" (Postill 2017).

Further still, nonhuman animals have a long history in and as media. As Berland (2019) points out, throughout recorded history, animals have mediated human relations and helped domesticate new technologies. For example, Berland argues that the prevalence of cat videos on the Internet help people become familiar with a vast, unwieldy, intimidating, and largely unknowable entity like the Internet. Finally, it is through the use of media technologies to communicate more ecologic values and long-term plans that people like Jared Diamond (2005) see the potential for humans achieving more sustainable lives.

Looking for affirmative worldbuilding projects in digital spaces is *not* a retreat to the techno-utopian discourse that surrounded the early days of the Internet. Jonathan Crary (2022) offers a recent and devastating critique of the hope that social media, or any form of digital media, can be an instrument of radical change. Crary makes a compelling argument that digital devices, processes, and networks are part of a global, capitalist apparatus whose outputs are the commodification of social lives, socioeconomic inequality, state-sponsored terrorism, and ecocide. In the aftermath of Crary's *scorched earth*, my search for affirmative projects seems a bit like finding a lone beetle in some postapocalytpic landscape and proclaiming, "See, there's still life. It's all going to be OK." However, the search for affirmative worldbuilding projects in digital spaces is more than a naïve insistence on hope. Searching for affirmative digital rhetorics is both a reclamation and a recognition. It is a reclamation in that it recovers the worthwhile communities, activities, and discourses that are taking shape on digital media. Capitalism and datafication may, indeed, make the digital age an uninhabitable spacetime, *and* it is worth remembering that well-crafted and heavily maintained apparatuses of capitalism and datafication are not the only ways to exist in digital space. There are other ways of being human and being networked that this book is dedicated to affirming. The search is also a recognition that, as Haraway (1985) would put it, humans are already cyborgs, and acting within digital spaces is simply living the social work maxim that work begins where the client—or person—*is*.

Although Donna Haraway (1985) pointed out decades ago that humans were already cyborgs, that realization has only become more apparent as work on augmented reality by industrial media giants has accelerated. Google, for example, is prototyping a new version of Google Glasses (that may, as you read this, already be available) that would augment grocery shopping to allow consumers to quickly find—for instance—gluten-free products, display language translations in real-time, and provide captioning (Google 2022). Many tech giants are working to create virtual metaverses. Also, the COVID-19 pandemic accelerated and expanded the use of video conferencing as a means of personal and organizational communication. And these are just the more recent, dramatic examples—let alone the ubiquity of smartphones for activities of daily living.

While Grant Bollmer (2019) is right that people cannot think *about* media from an outside perspective because humans are always *in* media (13), it has quickly become evident that people are always in *multi*media, including Earth's atmosphere and smart homes that connect people to the Internet of Things. Indeed, to talk of the more-than-human world is not only to talk about the nonhuman animals, plants, algae, bacteria, viruses, rivers, rocks, fungi, and mycorrhizal networks but also the server farms, satellites, algorithms, sensors, cameras, platforms, cryptocurrencies, and websites. Thus, I turn to worldbuilding rhetorics on digital media to see what kinds of flourishment and vulnerabilities are emerging in our physical and digital spaces. To do this, I need to briefly explicate how I am approaching the digital by highlighting the importance of the materiality and reality of media; exploring the performativity, multimodality, and immersive qualities of media; and addressing how media intervene in, as well as have a constitutive role in, cultural practices and relations.

My approach to media follows the work of material media theorists (Bollmer 2018, 2019), animal and environmental studies media theorists (Giraud 2019; Jue 2020), feminist media theorists (Berland 2009, 2019; Sharma 2017), Black media philosophy theorists (Towns 2022), and Indigenous media theorists (Miyarrka Media 2019; Todd 2015) in exploring media as pervasive, material, relational, embodied, and embedded in a more-than-human world. As previously mentioned, people do not think *about* media but think *in* media, and media abound with bodies—bodies of knowledges, bodies of practices, the configuration of bodies, the representation of bodies. And, as Towns (2022) writes, bodies are also media, and how one's body mediates the world and is mediated by the world is significant for epistemological and ethical claims. Much has been written about the perils of media—from algorithms of oppression (Noble 2018) and weapons of math destruction (O'Neil 2016)—and such perils are important sites of attention. Digital media provides many violent and destructive tools, yet it is hard to imagine human life without digital media (not only the Internet but something as mundane and unassuming as a microwave clock).

Still, it is possible to gain new perspectives on media environments that challenge taken-for-granted assumptions about what they are, how they operate, how they arrange human sensoria, and how they move bodies. For example, for some Aboriginal communities in Australia, media and media technologies are part of *Buku-manapanmirr*, a Yolŋu word meaning "joining together" (Miyarrka Media 2019, xvii). As a technology of *Buku-manapanmirr*, mobile phones are part of how community members join together with past and present peoples in mutuality and inclusion *with* distinction and the acknowledgment of different perspectives. Rather than using the phone as a distraction, the phone is part of the practice of present-ing people together.

Media are ripe with affordances and limitations. Media provide constraints on the realm of possibilities and make certain actions and values more probable, but they are not deterministic. For example, the medium of the ocean limits my capacity to breathe without training (e.g., Budimir Sobat can hold a single breath under water for over twenty-four minutes) or technological assistance, but sharks have thrived in this medium for hundreds of millions of years. Yet, other media—like the medium of pollution—alter the oceanic media in ways that render sharks vulnerable. Media afford and limit, and are afforded by and limited by, bodies in complex assemblages.

The same is true of digital media. Digital media are still material media consisting of the resources, technologies, labors, and relations that shape, energize, and sustain the distribution of audiovisual signal traffic (Parks and Starosielski 2015, 5) that act on and are acted on by other material bodies. And although these signals may be audiovisual, that does not mean that they do not act on the sensorium more broadly. Von Mossner (2017) describes how visual media are inherently multimodal: they activate somatosensory, emotion-related, and motor brain networks. Sound, meanwhile, can perform generative, locative, and comparative functions (Gallagher, Rosenfeld, and Tomlinson 2022). That these sensations are activated by digital stimuli does not detract from their embodied experience. A point made by both Bollmer (2018) and Chalmers (2022) is that online and offline are both real, but that does not mean there are not distinctions in how they mediate. Often, people exist in online and offline spaces, relations, and conversations simultaneously with attention shifting between the physical and the digital. This last point by Chalmers—the simultaneous realities of online and offline experience—is crucial to the significance of the digital worldbuilding rhetorics explored in this book, so I will spend a bit more time on it.

In *Reality+: Virtual Worlds and the Problems of Philosophy*, Chalmers (2022) explores how virtual worlds, including augmented realities technologies, act in physical worlds. Two aspects of Chalmers's analysis are particularly important for digital worldbuilding rhetorics: that virtual bodies are real bodies and, relatedly, that augmented reality creates multiple, real worlds. Concerning the reality of virtual bodies, Chalmers writes that a person's avatar in a virtual world serves as that person's locus of action in that virtual world, an avatar is a *real, virtual* body. Physical and virtual bodies are real and interdependent but nonidentical. The nonidentical part is where things can get messy, as when someone experiences a virtual body as one's physical body, what Chalmers refers to as an embodiment illusion. Chalmers gives the example of spending time in a virtual space with a tall(er) virtual body and then reaching for something in the physical space that is out of reach of one's physical body because the person's phenomenological experience of their own body was mediated by the experience of the tall(er) virtual body.

Chalmers also argues that augmented realities create multiple, real, virtual worlds and illustrates this with the image of a virtual world existing in a physical park through the use of augmented reality glasses. There may even be multiple pairs of glasses, supplied by different digital technology giants (e.g., Meta Platforms, Inc., Google, Apple). In the physical world there is no piano, in Google Reality there is no piano, and in Apple Reality there is a piano. This is not absolute relativism, Chalmers makes clear. There are still empirical truths about the presence or absence of a piano within these realities, or worlds. "We simply need to allow that relations are part of reality" (232).

Chalmers's book thinks through the difficult conceptual terrain that virtual realities present for life. I want to offer here, though, that from a media studies perspective, reality is always already relational and augmented—or, rather, mediated. The recognition of relationality and mediation is found in Indigenous epistemologies in which knowing is understood as knowing in relation to land (Arola 2018), and in materialist media theory, in which oceans (Peters 2015; Jue 2020), time (Peters 2015), and (more-than-)human bodies (Berland 2019; Towns 2022) are all understood as relational media. For example, Jakob von Uexküll's (2010) notion of *umwelt* considers how every lifeform has its own unique perceptual world and that perceptual matrix shapes that animal's experience of and actions in the world. The mating red-tailed hawk pair that nest around my home can see rodent urine all around because raptors can detect the ultraviolet light reflected by urine. The raptors' world is shaped by the presence of urine in ways my world is not. One could of course say that the rodent urine is present whether I see it or not, but the point here is how the perceptual capacity to detect rodent urine shapes how one moves through space. As Thom van Dooren writes (2019), "To insist that we make and inhabit *worlds* (in the plural) is not to deny a common space of existence. Rather, it is to hold such an idea in tension with another, namely, with the sense that each of us, individually and collectively, also crafts and inhabits distinctive spaces of existence" (8).

What Chalmers (2022) is arguing about (more-than-)human virtual worlds has always been true of (more-than-)human worlds, including the digital spaces of Internet with or without avatars. Smartphones have always mediated ways-of-being in the world. Think of the profound agency they exercise on food consumption, especially when someone travels. Various websites and apps first provide a filter by providing users knowledge of the existence of *some* eateries (i.e., not all eateries show up or show up easily) and then allowing users to select additional filters, such as average customer rating or price of meal. Of course, no one is really arguing that digital spaces do not impact commerce or, as Shosanna Zuboff documents (2019), even the nature of capitalism. Rather, what may be less apparent is how digital technologies and techniques are part of the ecology of the biosphere.

My launching point in this book is that digital worlds—even the mundane practices like scrolling through a Facebook or Instagram feed—also impact the (more-than-)human physical world. The physical impact refers to more than the obvious ways that the Internet requires substantial energy and the construction of massive server farms *but also* by the generation of worldbuilding rhetorics—through digital discourses and ways-of-mediating—that constitute the world in particular ways, reveal the recalcitrance of the physical world differently, attune senses to the (more-than-)human world in new ways, allow for the emergence of affective commonwealths, discipline through repeated procedures, and require and ignite different passions and energies.

The next three chapters of this book focus on and theorize three digital, worldbuilding capacities of rhetoric—disclosure, transformation, and infrastructuration—across four digital spaces (Instagram; Facebook; and the websites for the Pink Chicken Project and Reclaim the Void). Rather than trying to determine *how significant* digital rhetorics are to offline relations, I am more interested in illuminating *what kinds* of worldbuilding projects these digital rhetorics enact. How do these digital rhetorics constitute the more-than-human world? What materialities impact and are impacted by these digital rhetorics? How do these digital rhetorics engage the sensorium? What are the affective capacities and asignifying properties of these digital rhetorics? How do the procedures of digital rhetorics impact bodies? What (more-than-)human energies are involved in these digital rhetorics? How do digital rhetorics (re)attune people to the more-than-human world? In short,

I am interested in what digital media perform, or do, in more-than-human worlds. Attending to these questions means attending to the ways images circulate and act in media ecologies (Gries 2015), how digital design engages with people's sensate experience in the biosphere (Arola 2018), and how text is used not only to convey messages but how tags are used to catalog unfathomable amounts of data, or, as John Durham Peters (2015) says, to reduce the universe to an index (320).

Of course, the cataloging of information is never straightforward or pure; it is a worldbuilding activity itself that not only orders but acts and orients (Duncan 2022). But tagging is not only about indexing; it is also about *Bukumanapanmirr*—about joining together. As Jackson, Bailey, and Welles (2020) argue, hashtags are not only ways of cataloging conversations, they are forms of activism. Hashtags contribute to the efficacy of (counter)publics to join their politics, reach allies, build networks, and share narratives that are built both digitally and corporeally. As the #MeToo movement evinced, hashtags are fluid forms of joining together that traverse online and offline spheres with ease.[2]

Roopika Risam (2019) draws attention to the mobility that exists along the digital-physical continuum by making interventions into the digital cultural record part of pedagogy. Risam invites students to decolonize knowledge production by engaging in activities like digital mapmaking and editing Wikipedia. Engaging in these practices, students come to see themselves as participants in epistemology and can better critique—and intervene in—digital epistemological practices. Digital spaces, she notes, are capable of facilitating and foreclosing perspectives. In the final chapter of this book I discuss the implications of digital, worldbuilding rhetorics for pedagogy-and-practice, but for now, it is sufficient to register that digital rhetorics can and do participate in physical worldbuilding projects. And these worldbuilding projects are never neutral, which is why they are important.

STORY TIME: INVENTING A FLOCK, PART 1

In the fall of 2020, during the high time of pandemic-induced confinement, I read a letter in the *New York Times* that both fascinated and bewildered me. Ivan Kreilkamp (2020) described how raising chickens helped expand his emotional capacity by inviting him to think of interspecies relations not in terms of an anthropocentric understanding of *love* but more of a respectful recognition of otherness that can flourish among beings living together. This resonated with me.

He also described how his chickens were fundamentally a part of a flock, and he wondered if one could even care for an individual chicken, separate from the flock. Chickens, he claimed, resist being cared for in person-like ways because they are not singular; their identity is a communal identity. Knowing the objections some would have, Kreilkamp (2020) added, "To say this could seem to denigrate them, but that sense itself speaks to a problem: that so many of our standards of respect, dignity and value are linked to individuality and autonomy." I both understood and appreciated the attention taken to the importance of a *flock* for chicken wellbeing—that being a member of a flock is a foundational aspect of a chicken's way-of-life; however, I also was perplexed by both the idea of the flock as some stable entity and the loss of the identity of each individual making up this flock.

Flocks are always being invented, rearranged, performed. In May of 2021, we had our first of two experiences of thinking a chicken had been taken by a hawk, mourning the loss of that chicken's life, and then being startled by her reemergence many days later. We followed the hen and found that she was sitting on a brood of eggs in a tucked away and well-covered area. Given how long she had been missing, I knew the eggs could not be far from hatching.

On June 6, 2021, I heard some gentle peeps coming from the nesting area. I witnessed the Australorp—the hen who was hatching these eggs—fan out her neck feathers and make a rumbling sound when most other chickens approached. I approached her with some water in an eyedropper, as it was a hot day and I knew she would not leave the nesting area until the next day to give as many chicks as possible the opportunity to hatch before taking the brood out to begin foraging. She did not fan her feathers or rumble; she took in the water droplets as they came down her beak.

She remained calm, so I remained a few feet outside her enclosed area—watching and listening. One of my companion goats came over and sat against me, and one of the other hens came and sat atop the goat. The Australorp's feathers remained flat and her only noises were gentle chuts to the new arrivals. One of the ISA hens—a particular breed of chicken—approached, and the Australorp again fanned her feathers and rumbled. The ISA left, and her feathers relaxed back against her neck.

I became aware that although the flock shared the same coop and would respond similarly by sheltering together when the rooster let out a warning call, there were individuals with particular relations in this flock. The Dominique's presence was acceptable while the chicks were hatching, this ISA's was not.

The Australorp did not attempt to take the new hatchlings into the shared coop, but they were not safe without a coop. So, we built a second coop, which we nicknamed Coop 2. For a couple of months, the Australorp and the four chicks who hatched shared this coop. One night, about ten weeks later, I noticed the Australorp had returned to Coop 1 and the juveniles remained in Coop 2. Weeks more passed, and the two pullets and two cockerels joined the others in Coop 1. For a long time, Coop 2 remained empty, being used mostly as an additional egg-laying spot during the day.

Then, on October 3, 2022, I noticed Hera—one of the hens who came from that 2021 hatch—was missing from Coop 1. I feared that she either had been taken by another animal or that she was off nesting on a brood. I went to check Coop 2, and there she was, sitting alone on a roosting pole. The next night, she was there, again, but this time she was joined by Juno, the other hen from the 2021 brood. Weeks later, the two of them continue to stay overnight in Coop 2.

I cannot begin to speculate on why the two of them returned to the smaller coop, at least not without creating anthropomorphic explanations. I do know that they are not sitting on eggs in the second coop, there is enough room in Coop 1 for them, and they spend time with the other chickens during the day. They just sleep and spend their evenings alone in the second coop. I am also unable to offer any insight on how and why the departure began with Hera and then became Hera and Juno. The Australorp, Hera, and Juno reveal that flocks are in flux, comprised of individuals-in-communities engaged in their own worldbuilding projects.

NOTES

1. "Its function [*ergon*] is concerned with the sort of things we debate and for which we do not have [other] arts and among such listeners as are not able to see many things all together or to reason from a distant starting point. And we debate about things that seem to be capable of admitting two possibilities; for no one debates things incapable of being different either in the past or future or present, at least not if they suppose that to be the case; for there is nothing more [to say]" (Aristotle 1991, 41).

2. Part of the fluidity of joining together includes how foci shift. While the hashtag afforded a vast articulation of women's experiences, it also marginalized the pre-Twitter history of activist Tarana Burke's work in the #MeToo movement that predated the viral hashtag by over a decade (Ohlheiser 2021).

Chapter 4

Disclosure

The adoption of plant-based diets is one way the people engage in physical, political, and economic practices of building more flourishing more-than-human worlds. With increasing urban density, much of the Earth's population is not confronted with daily tasks of living well with nonhuman animals, outside of pets and the animals deemed *pests* with whom humans negotiate life and death.[1] Yet, humans are confronted with choosing what—or whom—they put on their plates each day. The adoption of a plant-based diet does not guarantee ethical eating (more on this later), but it is a choice that people are increasingly making for reasons ranging from personal to animal to environmental wellbeing.

Arguments for plant-based diets have existed throughout recorded history, from vegetarian diets (abstention from consuming animal meat) to veganism (abstention from consuming animal meat and animal products, including eggs, dairy, and honey) and lifestyle veganism (abstention from using products made from animals, like leather, or were tested on animals, like cosmetics). In Ancient Greece, Pythagoras urged vegetarianism based on a belief of transmigration—that is, the belief that human souls could pass into the bodies of animals; in East Asia, vegetarianism was part of an ascetic practice that was regarded as virtuous. Later, Leonardo da Vinci argued for vegetarianism out of compassion for nonhuman animals. In the 1800s, Anna Kingsford (2018), an antivivisectionist and animal rights advocate, combined arguments from animal ethics, human rights, human health, and biological sciences to recommend a vegetarian diet.

Starting in the mid-twentieth century, vegetarian and vegan rhetorics abounded. Carol Adams (2015) argued that the *sexual politics of meat*—that is, the way that both women and animals are erased and dominated by patriarchal practices—made vegetarianism a feminist practice; Peter Singer (2009) proclaimed vegetarianism a liberatory practice because it allowed the most happiness and least suffering to the most beings; and Tom Regan (2004) made a case that vegetarianism was part of animals' duties based on the moral duties owed to sentient beings. In the early twenty-first century, Syl and Aph Ko (2017) were among the intersectional and Black vegan scholars to argue that veganism is a liberatory practice for people as well as nonhuman animals and that the construction and maintenance of the human-animal divide is what allows an Eurocentric ideal human to be positioned on one side of that divide and nonhuman animals on the other, with human Others (people who are marginalized) to be positioned somewhere between (66). Veganism is a dietary practice that disavows a human/animal divide.

Vegetarianism and veganism—as a social justice praxis that seek to undermine systems oppressive to (non)human animals—may be liberatory practices for people who have been marginalized, *and* they are also marginalized identities. Historically in Western culture, vegetarian and vegan identities have been marked by caricature and exclusion, subject to mockery and derision. In 2018, the editor of *Waitrose & Partners* food magazine responded to a proposal for a plant-based issue by saying that he was in favor of an issue devoted entirely on how to hunt and eat vegetarians/vegans (Aguilera-Carnerero and Carretero-González 2021). Commercials for popular fast-food chains have been grounded in anti-vegetarian humor, such as Hardees' "Save the Veggies" and Arby's' "Friends Don't Let Friends Eat Tofu." Beyond these more flamboyant degradations, vegetarians and vegans also contend with everyday exclusions, such as being barred from participating in a particular event because of limited food choices or being unable to interact freely with others because of the perceived need to avoid letting others know that one is vegetarian/vegan (Buttny and Kinefuchi 2020; Paxman 2021; Romo and Donovan-Kicken 2012).

In recent years, vegetarianism and veganism have become more mainstream. In 2004, there were just under 300,000 vegetarians and vegans in the United States. By 2019, that number had climbed to 9,700,000—a 3,000 percent increase in fifteen years (Glusac 2022). Further, African Americans make up the fastest-growing vegan demographic in the United States—8 percent of African Americans identify as vegan, which is nearly three times greater than Americans overall (McQuirter 2021)—and have created the Black Vegan movement and #BlackVegan digital activism. During this time frame, there also have been highly visible celebrity endorsements of vegetarian and vegan diets, from Moby's TEDx talk to Joaquin Phoenix's Oscar speech to Jackie Chan and Arnold Schwarzenegger's *The Game Changers* to Rooney Mara's founding of the vegan clothing line Hiraeth. Competitors on popular baking and cooking competitions have brought their vegan recipes to the competition (e.g., Freya Cox in *The Great British Bake Off*) and left the competition when tasked with cooking meat (e.g., Alex Lenghel in *Romanian MasterChef*). In the United Kingdom, many comedians who are vegan have raised the profile of the diet by discussing that aspect of their identity in humorous ways on popular panel shows, and prominent vegetarian and vegan athletes—including Fauja Singh, the world's first centenarian marathoner—have helped to dispel misconceptions about vegetarianism and veganism being related to physical weakness. Perhaps the final frontier in plant-based ubiquity has been the shift of fast-food chains to actively courting plant-based dieters.

Concurrent with the marketing shift in the fast-food industry, the past fifteen years also have been marked by (1) a proliferation of plant-based meat substitutes and (2) the growth of social media platforms. "Plant-based meat alternatives" (PBMAs) refer to products that have the aesthetic qualities (e.g., taste, smell, appearance) of traditional meat products but are composed entirely of plant products. PBMAs differ from other vegetable dishes in that the goal is not simply to replace a meat with a vegetable dish but to replicate the qualities of meat with plants. For example, a black bean patty or a "meatless ground" (PBMAs made to have the qualities of ground beef) could both be used to make a vegetarian burger, but the PBMAs is made to replicate the sensory experience of eating an animal-based meat product. Beyond Meat—an industry leader in PBMAs—founder Ethan Brown said his mantra is for people to eat "more of what they love" while eating fewer animal products (Darmiento 2020).

PBMAs are being sold in major grocery chains and adopted on the menus of major chain restaurants, including Pizza Hut, Burger King, Denny's, and TGI Friday's. The widespread adoption of PBMAs not only make them accessible alternatives to meat but are also part of their economic success. PBMAs are now a multibillion-dollar industry, and it is projected that PBMAs will be worth $31.4 billion by 2026 (Rodrigues 2022). PBMAs have been hailed for offering new ways of eating plant-based diets and broadening the appeal of plant-based diets to people seeking vegetarianism, and they have alsobeen criticized for perpetuating ideologies that associate meat with masculinity.

PBMAs have allowed for new kinds of vegetarian eating, and the rise of social media platforms have allowed for new ways to share these foods and disclose vegetarian and vegan identities. Vegetarians and vegans have developed a variety of techniques for managing their identities, many of which have involved concealing or downplaying one's vegetarian or vegan identity, what some refer to as "self-silencing" (Bolderdijk and Cornelissen 2022). For example, vegetarians and vegans may avoid using the word "vegetarian" to describe their diet or provide more socially acceptable justifications for not eating meat, such as abstaining for weight loss. Techniques of concealment are employed in an attempt to avoid a confrontation with a meat-eating colleague, friend, or relative. On the other hand, omnivores often assume they will be judged negatively by vegans and vegetarians, which invites vegans and vegetarians to avoid engaging in behaviors and saying things that imply judgment (Romo and Donovan-Kicken 2012). Traditionally, vegetarianism and veganism have been associated with the animal rights movement (Phelps 2007), and declarations of vegetarian and vegan identity are assumed to be associated with a moral condemnation of non–vegetarians and vegans. Hence, avoiding the word "vegetarian" and saying one does "not eat meat" in its place evokes fewer political and ideological assumptions attached to words like "vegetarian" and "vegan."

In contrast to these techniques of concealment in face-to-face interactions, Estok (2021) describes how social media platforms have enabled making one's dietary identity highly visible. On the various platforms, contributors are able to "take control" of their identities and tell stories, talk about food choices, share recipes, and post pictures (335). In contrast to mainstream animal rights visual campaigns (e.g., People for the Ethical Treatment of Animals' "I'd Rather Go Naked Than Wear Fur"; the release of undercover footage at factory farms), social media platforms enable—as well as constrain—the proliferation of *vernacular plant-based visual rhetorics*.

Vernacular plant-based visual rhetorics is the term I am using to refer to the rhetorical culture of plant-based diets expressed in/through the quotidian imagery of social media; these digital rhetorics are significant owing to (a) their capacity to afford a readiness to disclose one's participation in plant-based diets and (b) their desirable depiction of plant-based diets, which can afford a willingness to try vegetarian cuisine (more on these points later in the chapter).[2] As quotidian practices, *vernacular plant-based visual rhetorics* provide access not only to the headline-grabbing efforts of activists (which is also a rich and important discourse of study) but to the local knowledges, concerns, values, logics, aesthetics, and traditions shared among people who are making plant-based diets visible through digital food practices.

People interpret images, whatever the subject matter, through *image vernaculars* (Finnegan 2005)—the relatively stable, culturally and historically situated grounds through which people make sense of an image—and the plant-based visual rhetorics work to (at least partially) destabilize food or dietary image vernaculars typically associated with carnivore diets. As Gerard Hauser (2019, 14) might put it, whatever else these images are, they are inducements to an attitude, or—as I will demonstrate shortly—at least inducements to sensations that afford the possibility of an attitude shift.

The digital food practices of *vernacular plant-based visual rhetorics* are part of the food photography phenomenon, in which individuals post pictures of their meals on their social media accounts. Digital food photography, Lewis (2020) asserts, works against the more sweeping generalities of dataveillance (i.e., the monitoring and collecting of online data, such as credit card transactions and GPS location) by allowing contributors to diarize their lives in networked and public ways that map and capture the intricacies of one's food practices.

The networked aspect of digital food practices may be particularly important for vegetarian and vegan identities because they offer people the opportunity to find allies, to confirm "I am not alone." For example, in a face-to-face setting, Bolderdijk and Cornelissen (2022) found that self-silencing can be mitigated by the presence of pro-vegetarian or vegan sentiment. Results of their experiment indicated that people were more likely to sign a petition calling for plant-based food options in a room full of people ridiculing the petition if there was an ally in the group who indicated support for the petition. With their sheer abundance of users and posts, social media platforms offer the opportunity to connect with allies. However, locating allies on social media requires sophisticated cataloging protocols to find posts of relevance in the overwhelming volume of tweets (there are five hundred million tweets on Twitter each day, Smith 2020)—and Instagram (with its archive of over fifty billion photos uploaded as of 2022, Omnicore 2022). Hashtags, a word or phrase preceded by the pound symbol, help users and algorithms to sift through the masses and find posts that are part of a conversation of interest.

Recently, scholars have looked at the power of hashtags to network and amplify conversations on- and offline. Hashtag activism (Dadas 2017; de Kosnik and Feldman 2019; Jackson, Bailey, and Welles 2020) refers to the use of hashtags to engage in civic debate and social movements. Among the more visible acts of hashtag activism in recent years has been the #MeToo movement, which also generated related hashtag discourse communities like #NotAllMen and #YesAllWomen. The #MeToo movement was not only a robust digital discourse but also a central story in mass media and contributed to the adoption of new policies on sexual harassment and the firings or resignations of many people who had been reported to be involved in sexual harassment. Not all instantiations of hashtag activism are associated with such widescale outcomes. However, the presence of communities afforded by hashtag networking can still contribute to what Chris Ingraham (2020) calls an "affective commonwealth," or a sense of sharing what it feels like to live a life (5).

Plant-based hashtags do not have to be as visible as #MeToo to create pro–vegetarian and vegan affective commonwealths that could be helpful in enabling people to perform rather than silence their vegetarian and vegan identities. Certainly, the presence of allies can help buffer against the substantial anti–vegetarian and vegan rhetoric that also circulates in online discourse communities. Even some pro-plant-based discourse, such as the popular documentary *The Game Changers*, tries to distance itself from vegetarianism and veganism by avoiding those terms in favor of *plant-based*. Thus, surveying an affective commonwealth animated around plant-based diets can be a challenge. The labels can gesture to important rhetorical ruptures under the broad term *plant-based*, but the labels do not provide much information on what practicing "vegan" or "plant-based" actually looks like.

In this chapter, I use techniques of visual content analysis and visual rhetorical criticism to understand better what lines of continuity and distinction *show up* in people's presentation of plant-based diets. This chapter looks at multiple plant-based diet hashtags and explores how and with what consequences vernacular plant-based visual rhetorics show up on Instagram—that is, it examines the characteristics of the images and descriptions people are cataloging with popular vegetarian, vegan, and other plant-based hashtags.[3]

Both Lewis (2020) and Estok (2021) acknowledge the importance of visual content to digital food practices, and scholars, and researchers such as Erben and Balaban-Sali (2016) have noted the substantial role that social media play in contemporary vegan and vegetarian identities and activities, but there has not yet been a visual content analysis of the posts of Instagram users who use vegetarian and vegan hashtags. Vegan ideology has been studied through visual analysis of graffiti (Hughes 2021), but not much attention has been devoted to the day-to-day performance of plant-based identities and practices through social media posts.

In this chapter, I take this journey by first elaborating and bringing together theoretical understandings of *disclosure* from interpersonal communication and rhetorical theory perspectives and introduce *dietary image vernaculars* and *meat semiotics* as important aspects of plant-based diet disclosure. Then, I analyze a selection of Instagram images with plant-based hashtags in three parts: (1) how and to what extent these images participate in creating plant-based diets that are indistinct from meat diets, (2) how the production of these images afford feeling closer to familial and familiar dietary practices, and (3) how and to what extent these images constrain more-than-human worldbuilding through the rhetoric of "no sacrifice." This chapter illuminates how particular forms of flourishment are intertwined with particular vulnerabilities—in this case, how the inclusiveness afforded by PBMAs also allows for the (partial) assimilation of plant-based diets into hegemonic meat culture. The tension between inclusion and assimilation invites sustained attention to how a diet—or any social or environmental justice practice—shows up in the world.

DISCLOSING PLANT-BASED MEALS

Earlier in this chapter, I touched on how people eating plant-based diets disclose their identity and/or dietary practices in human relationships. These disclosures mostly occur in one-to-one and one-to-small group face-to-face situations. With the focus on vernacular plant-based visual rhetorics, I am looking at one-to-*n* disclosures in digital spaces, and I use the term disclosure in a way that takes into account two approaches to the term, one from interpersonal communication theory and one from rhetorical theory. Thus, before I offer a working understanding of how I am using the term disclosure, I need to describe briefly how the term shows up in these two different traditions, as well as provide a brief accounting of the role Peircean semiotics plays in disclosure.

Self-disclosure is an important concept in interpersonal communication, and one that is particularly relevant to a book exploring the entwinement of vulnerability and flourishment. Self-disclosure, as distinguished from *disclosure*, is characterized as an intentional act of honest revealing about oneself to others. Interpersonal communication scholarship describes self-disclosure as a hallmark of relational intimacy and satisfaction; it is both what makes one vulnerable in a relationship (e.g., subject to rejection, the consequences of negative impressions, the possibility of hurting relational partners) and also what allows one to flourish as the person one is/is becoming (e.g., able to receive self-validation, engage in self-clarification, participate in relationship maintenance). The act of disclosing can also be cathartic, even when the recipient of disclosure is unable to respond to the content of a disclosure, such as in human-animal communication (Craig, Nieforth, and Rosenfeld 2020). Theories of self-disclosure, like Petronio's (2013) Communication Privacy Management Theory, offer explanatory frameworks for the centrality of self-disclosure in all interpersonal communication (e.g., in intimate partner communication, family communication, organizational communication) and how self-disclosure is not only an act (i.e., the utterance) but also a nonlinear process (i.e., disclosure does not start at a non-disclosive starting point and progress toward some totally disclosive end) that unfolds in complicated situations and relations.

Self-disclosure—the sharing of a person's biography as well as one's thoughts, attitudes, and emotion—can increase affiliation, and this affiliative capacity has also been found through disclosure on social media platforms, in which disclosure allows people to recognize commonalities in experiences or beliefs (Ledbetter et al. 2011) and to experience catharsis through sharing (Vilhauer 2009). As with all things in this book, the preceding statement should not be read as a straightforward *positive* consequence. It is a consequence that participates in worldbuilding through relationship-building, but the acknowledgment that it builds relations alone tells us nothing of the architecture or construction materials of this relationship. In other words, the rhetorical capacity of self-disclosure to build affiliation does not provide any information about the ethics of the affiliation at hand. As with all worldbuilding capacities, there is always the dual task of recognizing what relations, entities, and subjectivities are brought into being and how *and* also of analyzing the ethical nature of those relations, entities, and subjectivities. I will return to the ethics of relation-building and worldbuilding in a moment, but first, let me broaden the scope of disclosure under consideration.

Beyond self-disclosure, there is a wider matrix of disclosures, in which intentional communication is only one possible way that an entity comes to be known. This is the rhetorical understanding of disclosure, as theorized by Thomas Rickert (2013) through his reading of the philosophy of Martin Heidegger, in which the world and everything in it is disclosed and concealed through modalities (i.e., a mode in which something is encountered, experienced, or expressed). Visual and auditory modalities, for example, are two ways that the world gets disclosed to humans—that is, are two ways the world is made available or accessible—but those modalities are always partial. Disclosure in this sense is an ongoing, situated, (more-than-)human process by/through which things *show up* in particular ways. To say the (more-than-)human world is disclosed through multimodalities is not to say that these modalities are disclosed to *tabula rasa* perceivers. Beings are attuned to the world in different ways. As Rickert says, there is no state of being unattuned—or no blank slate from which one can be "correctly" tuned—to the (more-than-)human world; there is only changing the tuning (162).

Further, there can be competing attunements and disclosures, even in a (seemingly) single encounter. For example, the smell of a Beyond Burger discloses an aroma of animal meat; yet, the visual encounter of symbols on the package disclose the patty as a complex configuration of water; oils and butters; pea, rice, and mung bean proteins; potato starch; apple extract; lemon juice; and other ingredients. Neither modality is exhaustive but each reveals different aspects of the plant-based burger. "The arrogance of human beings comes from forgetting this, from equating our knowledge and practical handling of something . . . with what that thing *is*" (Rickert 2013, 212).

The epistemic humility warranted by Rickert's (2013) caution of human arrogance is closely connected to the ethics of disclosure. Because disclosure—including self-disclosure—is an ongoing process, beings must remain open to new disclosures and the possibilities of relationship remaking. For Rickert—and for me—activities that foreclose the possibility of future disclosures are ethically compromised because they deny the unfolding nature of (more-than-)human flourishment.

I can now bring plant-based diet disclosures back into the conversation to help expand on this point. On a personal level, approaching a relationship from an ethics of care perspective calls for openness to a person's becoming. This can be challenging in friend and family situations in which historically and experientially informed schemas have been created. Although a plant-based diet may sound like a small disclosure to make, it is articulated with significant relational implications, for example, like a feeling of moral criticism of others by the person adopting a plant-based diet or the interference in or alteration of relational patterns (e.g., provoking a necessity to abandon a favorite hangout that is not friendly to the new diet, foregoing a traditional dish that is of significance to one's family and culture).

It is on the latter point—the interference in relationships—that PBMAs offer a way of disclosing plant-based diets differently. Instead of the caricaturized image of a vegetarian or vegan who only eats kale, the proliferation and distribution of PBMAs afford the opportunity for people adopting plant-based diets to participate in many relational and cultural practices. As will be demonstrated, PBMAs allow individuals to participate in plant-based diets while also caring about culinary identity markers. While PBMAs do not alleviate the tensions around the moral critique that some feel are implicit in one's adoption of a plant-based diet, PBMAs do disclose the diet—through sight, smell, and taste—in remarkably familiar ways. This capacity is perhaps most evident in the food photography of social media, in which plant-based diets can be disclosed differently—sometimes completely indistinguishable from images of omnivorous diets. In the following analysis, I address the affordances and constraints of such visual disclosures, but first, I need to provide a brief accounting of the roles *dietary image vernaculars* and *meat semiotics* play in disclosing plant-based diets differently.

DIETARY IMAGE VERNACULARS AND MEAT SEMIOTICS

Cara Finnegan (2005) introduced the term *image vernaculars* to describe how images can function rhetorically as enthymematic arguments. Image vernaculars are the colloquial, everyday ways that people encounter and make sense of images. For an image to function as a vernacular argument, the image must work with everyday, enthymematic modes of reasoning that are based in social and cultural norms in context. Enthymemes, as argumentative devices, are a form of a syllogistic reasoning that truncates or suppresses a premise, which is then constructed by the interpreters using their tacit knowledge. Because the interpreter is actively involved in the argument's construction, the enthymeme is a powerful rhetorical device that seems *natural*, rather than context-bound and constructed.

Dietary image vernaculars, as I am using the term, are the colloquial, rhetorical grounds through which people make, share, and encounter food photography and food displays. *Vernacular plant-based visual rhetorics* featuring PBMAs intervene in dietary image vernaculars by disrupting the enthymeme. If a person shares an image of a patty accompanied by tomato, lettuce, and onion and sandwiched between buns, a culturally and historically situated line of reasoning would lend one to conclude that this person eats animal meat, but PBMAs introduce ambiguity to this line of reasoning. This may sound like a trivial example. A hypothetical person saw an image of a burger on a friend or colleague's social media and wrongly concluded that the person ate meat—*so what*?

Humans frequently use dietary image vernaculars, not only photographs but also in-person observations, to take account of one another's dietary preferences and needs. Sharing food is often an important part of diplomacy, of cultural exchange, and in communicating care. A surprise meal is a common enactment of care among friends, colleagues, and lovers. When dietary image vernaculars are disrupted, through the existence of plant-based meats or vegan cheeses, new affordances and constraints regarding dietary disclosure are introduced (explored in more detail below). And this raises the question of *how* meat semiotics are altered and the mechanism by which vernacular plant-based visual rhetorics intervene in dietary image vernaculars.

Semiotics refers to the theory of signs and their interpretation—and in this chapter I focus on Peircean semiotics. For Charles Sanders Peirce (1998), semiotics was central to his philosophy of pragmatism and was the metatheory from which the polymath understood other areas of study. Peirce once wrote, "It has never been in my power to study anything,—mathematics, ethics, metaphysics, gravitation, thermodynamics, optics, chemistry, comparative anatomy, astronomy, psychology, phonetics, economics, the history of science, whist, men and women, wine, metrology, except as a study of semiotics" (Peirce 1977, 85–86). However, I take the position that semiotics—the account of signification—is one of the ways the (more-than-)human world gets disclosed; affect—an asignifying rhetorical encounter (Muckelbauer 2021)—is another way the world gets disclosed.

This chapter's engagement with Peircean semiotics is not the first time semiotics have been taken up to explore visual phenomena. Gunther Kress and Theo van Leeuwen (2021) use Barthes's semiotics to provide a complex visual grammar and, as I will discuss more in the analysis below, Carol Adams (2015) argued for the oppressiveness of Western meat culture based partly on the idea that animals were *absent referents*. Peircean semiotics is the *right* way to study the disclosiveness of vernacular plant-based visual rhetorics; it is *a way* that yields productive insights. One of the reasons Peircean semiotics fits well with the ethos of this book is that, as a theory of signs based in pragmatism, a sign's "meaning" is not understood to be static or singular. "Meaning" is only accessed through the consequences of a sign in the world (also as the *pragmatics* of signs) and those consequences are multiple and mutable. This may explain the appeal of Peircean semiotics to more-than-human anthropology and to the development of a field of study of the signs of life, known as *biosemiotics*.

What Peircean semiotics offer to the understanding of PBMAs in a community-care paradigm is a method for understanding how plant-based diet imagery can circulate through multiple, even conflicting, communities simultaneously. Some signs are, in the words of Sara Ahmed (2004, 90) "sticky," and allow many concepts, values, and practices to attach to them. Peircean semiotics help illuminate some of the adhesive properties.

In Peircean semiotics, the (more-than-)human world is replete with signs, and Peirce divided signs into three categories: icon, index, and symbol. An icon is a sign that has a physical resemblance to what is being signified; for example, a photograph of a Beyond Burger is an iconic sign of the physical burger. An index has a physical relationship to the signified; the grease stain left on a paper wrapper is an indexical sign of the burger. A symbol has neither a physical resemblance nor relationship to the signified and is a sign that must be culturally learned; the words "Beyond Meat" do not attach to the burger through resemblance or causal relationship but through the social and cultural attachment of meaning made possible by the alphabet.

In Peircean semiotics, all three categories of signs participate in semiotic relationships between interpretant (or sense), representamen (or sign vehicle), and object (or referent). The representamen, or sign vehicle, refers to the *signifying element* of a sign and not the sign as a whole; for example, if someone reads the words "Beyond Meat," the signifying element of this symbol is the combination of letters B-e-y-o-n-d_M-e-a-t and *not* the fact that someone is reading the sign in, say, black ink. This is not to say that fonts and colors do not matter or do not inflect the encounter of symbols in particular ways, but rather that the combination of letters for "Beyond Meat" would continue to attach to the plant-based burgers—through sociocultural learning—through the different fonts and colors. The object, or referent, places constraints on the signifying relationship by imposing certain parameters for a sign to represent that referent. To use the example of the index, an index must indicate a physical relation between it and the referent to be a sign of that referent. If I consume a Beyond Burger and there is no crumb or grease residue remaining on my plate, there is no visible, indexical sign of the burger. There may, however, still exist chemical indices—such as odors—that are perceptible to, for example, my canine companions. The interpretant, or sense, is the translation of the sign/object relation by focusing attention on certain features of the signifying relationship. The way that the presence of grease generates an indexical sign of the referent, the Beyond Burger, is by focusing attention on the physical connection between grease and burger. As already indicated by the example of Beyond Burger grease, Peircean signification is not limited to intention or to human communication systems. Signs abound in all sensorial modalities, but it is important to remember that the world is not directly accessed through their signs; the world is mediated—and disclosed—through signs that are themselves situated socially and in relations of power. The examples I just gave using the Beyond Burger to explain semiotic concepts were fairly simple. As I will show momentarily, plant-based diets are disclosed in social media posts in complex ways that occasion both consequences of (more-than-)human flourishment and are productive of particular (more-than-)human vulnerabilities.

A final point worth noting here before proceeding into the analysis is how these vernacular plant-based visual rhetorics defied my expectations. I looked through hundreds of images for each of several hashtags (#BeyondMeat, #ImpossibleBurger, #BlackVegan, #Vegan, #Vegetarian, #PlantBased, #BullCityVeganChallenge [a hashtag associated with a challenge to chefs in Durham, North Carolina, to create the best vegan dishes in the city], and #VeganChefChallenge [a hashtag associated with the nationwide expansion of the Bull City Vegan Challenge]) and engaged in detailed coding for over two hundred photographs representing a sample of those hashtags. Based on my readings of vegan, vegetarian, and plant-based rhetorics—including the histories of the terms, their association with different kinds of advertisements, their affiliation with different ethical orientations toward the diets—I expected images of vegan meals to have marked contrast from the plant-based, vegetarian, and plant-based meat alternatives hashtags. I *thought* this chapter would be about how vegan visual rhetorics disclose plant-based diets differently from other plant-based visual rhetorics and anticipated a visual rupture akin to the distinction found in the selection of labels. Instead of marked contrast, I found remarkable similarity. The way PBMAs show up across plant-based visual discourse communities is indicative of how PBMAs have changed what it means to eat a plant-based diet. Plant-based diets now afford greater inclusion in familial and cultural activities and with that inclusion comes the vulnerability of plant-based dieting being assimilated into meat culture.

FLOURISHMENT THROUGH SEMIOTIC ABUNDANCE AND CHANGING VERNACULARS

The 2016 photos of #BullCityVeganChallenge—a chef challenge that began in Durham, North Carolina, in 2010—features pictures of meals made with tofu and tempeh, pizzas with vegetables and vegan cheese, and a variety of desserts. In the 2021–2022 posts of the now-nationwide #VeganChefChallenge, photos included Beyond Chicken sandwiches, Korean short rib bowls, and Beyond Brats. #BeyondMeat was used to tag images of PBMAs used in rigatoni and baked feta pasta, tacos, Tim Horton's breakfast sandwiches, smoky Carolina BBQ, pork belly fries, and many burgers and fries. The text of one post in Spanish read that the burger was "sin carne . . . esta bomba" ("without meat . . . is [the] bomb"), a Norwegian post urged "hva med en deilig" Beyond Burger ("how about a delicious" Beyond Burger), and a Croatian post proclaimed there to be "*nema razlike*" ("no difference") in the PBMAs and animal meat burgers. #ImpossibleBurger posts featured chipotle vegan burgers with sweet potato fries, multiple-pattied burgers that hands could hardly hold onto, and stuffed cabbage paired with a pinot grigio rosé for a candlelit dinner.

The plant-based scene had changed a bit since the launching of Beyond Meat in 2012 and its Beyond Burger in 2016, and the arrival of the Impossible Burger in 2016. In one #BullCityVegan post from 2021, it is noted that in the Triangle area of North Carolina (Raleigh, Durham, and Chapel Hill and the surrounding communities) there were no vegan restaurants at the time the challenge began in 2010, and by 2021, there were eleven. The post notes the ubiquity of "vegan" and "plant-based food" in 2021. This ubiquity is partly afforded by the PBMAs that can be found in eighty countries (Beyond products) and eateries ranging from concessions at baseball stadiums to the drive-thru at fast-food restaurants to Michelin star restaurants like Harwood Arms in London.

In this section, I discuss how PBMAs allow plant-based meals to be disclosed differently, affording opportunities for different plant-based identities to take shape—as seen across Instagram posts using different plant-based hashtags. However, I want to be careful here. To say that I found PBMAs across plant-based hashtags is *not* to claim that they show up equally across hashtags. For example, #BlackVegan had more images of meals that were clearly plant-based, that is, there were more photos of food without PBMAs. This may be in part attributable to Black veganism's intersectional emphasis on veganism not only for animal wellbeing or environmental preservation but also veganism for human health. Black veganism approaches veganism as a liberatory practice in terms of social justice (prioritizing Black health as an aspect of social justice, as not participating in oppressive agricultural practices) *and* of personal wellbeing (Adewale 2021; Harper 2020).[4]

Thus, while no hashtag was without PBMA images, their distribution varied. Certainly, this is an aspect of *vernacular plant-based visual rhetorics* worth greater attention than I can direct at this time, as I work to first sketch what the presence of PBMAs does in/to plant-based dietary practices. In other words, I am now looking at what happens when PBMAs show up across all these hashtags—future work might address the density and significance of

PBMAs in different network configurations. For now, I begin by discussing the flourishment that PBMAs offer through their troubling of meat semiotics and how this troubling affords an ambiguous space to disclose a plant-based meal, diet, or identity differently on social media. Two important aspects of the way PBMAs trouble meat semiotics is through their *nema razlike* characteristics and their ability to mobilize social connectivity.

Nema Razlike!

In one Instagram post (identity of account withheld), a white, ceramic bowl—presumably containing ketchup or some other kind of sauce—sits next to a helping of golden fries. Some of the fries have a consistent wheat hue to them, others are burnt to a deep brown on the edges. On the other side of the fries is a sesame seed, wheat bun atop a white cheese oozing over the sides of a red-brown patty. The meal is, clearly, the recognizable *burger and fries*. The accompanying text is written in Croatian, and the text lists the ingredients and says the meal was air-fried before proclaiming, "*Nema razlike!*" *Nema razlike* translates to English as "no difference." Of course, there is a difference—a big difference. The cheeseburger is made with a Beyond Burger patty, which means the patty has no animal products. However, what the post is getting at is this new generation of "veggie burgers" (this is not even a term used by Beyond Meat and Impossible Foods; Impossible's label says "burger patties made from plants" and Beyond says "plant-based patties") does not have the sensorial qualities of a black bean mixture shaped into a patty but is instead indistinct from the animal-meat patties so often found couched between cheese and onions and a pair of buns. (Note: for ease of readability, I am focusing on PBMAs for beef burgers, but popular PBMAs include chicken, pork, and fish alternatives.)

This is, of course, the primary mission of Beyond Meat and Impossible Foods. Beyond Meat's short-form mission statement is "to create delicious, nutritious, sustainable protein so that you can Eat What You Love™, no sacrifice required" (Beyond Meat 2022), and Impossible Foods—with the tagline "meat made from plants"—say they exist to "save meat, and earth" (Impossible Foods 2022). A quick Google search for "plant-based burger taste test" videos returns over four hundred thousand results, and the first several pages of results feature various challenges: some done by vegans, some done by omnivores, some by adults, some by children. Many of the first fifty videos—except the tests done by vegans—are challenges comparing Beyond, Impossible, and beef and done without the participants knowing which burger is which. People often have a hard time knowing which burgers are the PBMAs and which ones are the beef, and sometimes they prefer the PBMAs.

On Instagram, multiple posts attested to the modal qualities of PBMAs being "so similar to the real thing" that they could not tell they were eating a PBMA. Both Beyond Meat and Impossible Foods market on their ability to provide PBMAs that allow people to eat plant-based diets with "no sacrifice." Later, I discuss some concerns with the "no sacrifice" rhetoric, but first, it is important to recognize that PBMAs have certainly and profoundly altered the context of plant-based diets by offering products that are *nema razlike* than animal meat. PBMAs disclose plant-based differently by intervening in meat semiotics.

Earlier, I brought up modalities and semiotics as key terms for understanding disclosure. PBMAs disclose plant-based meals differently because they trouble the association between multimodalities of the burger and the indices of the burger. The modalities of the burger are many and include the burger's smell, tastes, textures, the sound it makes while cooking, the color of the burger raw and once cooked. Many of these modalities have also, historically, been intimately tied to indices of beef: the smell of a burger cooking was recognizable as a sign of beef being cooked and the taste left in the mouth was distinguishable as a trace of ingested beef. However, PBMAs create an abundance in meat semiotics because they multiply the possibilities of the referent of these indices. No longer are these modalities and indices tied exclusively to an animal meat referent.

In the photograph accompanying the "*nema razlike*" text, the PBMA comes with cheese, and it is unclear whether the cheese is dairy or vegan, which is another aspect of ambiguity in vernacular plant-based visual rhetorics. The burger is paired with fries, and a "burger and fries" is a recognizable pairing that circulates through much of American and Americanized culture—from fast-food restaurants to bar food to cookout mainstays. Although the word *burger* does not, linguistically, gesture toward *beef* or *cow*, the dietary vernacular—linguistic and image—discloses a burger as a beef patty. Yet, the caption introduces doubt about what is being perceived in the image.

Vernacular plant-based visual rhetorics' *imitatio,* or imitation, of meat aesthetics and dislodging of meat semiotics problematizes dietary (image) vernaculars. This problematization is not confined to burger and fries. When PBMAs are photographed and shared on social media, they are often depicted in compositions that would, typically, indicate animal meat. People shared pictures of PBMAs serving as the ground meat in a pasta sauce or as meatballs topping a bowl of spaghetti, as the spiced meat covering the bottom of a taco shell, as the brat smothered in toppings in a hot dog bun, as the meat accompanying egg and cheese and tucked into a Tim Horton's wrapper, as a pizza topping, and as part of Korean BBQ. Whereas photographs of non-PBMA plant-based meals often showed isolated bananas, mushrooms, green beans, or other foods, PBMAs were shown in compositions in which they—as meat made from plants—defied expectations.

Nema razlike gestures to the increasing difficulty that PBMAs have introduced into deciphering meat semiotics, into the ways PBMAs have problematized dietary image vernaculars. Recall that in Peircean semiotics, indices—as well as icons and symbols—participate in a semiotic relationship between the interpretant (or sense), the representamen (or sign vehicle), and the object (or referent). The sign vehicle is that signifying element of a sign. Smoke may have a metonymic relationship to a cookout, but smoke is not a signifying element of a burger being grilled. Smoke is a signifying element of a fire. The smell of the burger wafting through the air, however, has long been a signifying element of beef.[5] However, some PBMAs—like Beyond Burgers—have been able to capture the aroma profile of the compounds 1-octen-3-ol, octanal and nonanal that provide the meaty, fatty, grilled characteristics associated with beef burgers. The consequence of this aroma-matching is that a prominent modality of beef burgers—the smell—is no longer strictly an index of beef cooking.

PBMAs interfere in the ability to infer what is being cooked. The smell of beef cooking may actually be the smell of a very particular composition of plant-based ingredients. Of course, it is not only in the aroma profile that PBMAs disclose plant-based diets differently. The *nema razlike* qualities of PBMAs mean that the multimodal qualities (e.g., the smell of beef, the taste of beef, the sizzle of beef) that once revealed aspects of beef—but, again, never disclosed beef *as such*—are mimicked well by PBMAs. In contrast, veggie burgers—from portobello mushroom caps to black bean patties—do not disrupt meat semiotics; the modalities through which veggie burgers are disclosed do not mimic, and therefore do not confuse or trouble, the indices of beef patties.

PBMAs disclose meat differently and enable different dietary disclosures, causing a semiotic rupture in our understanding of "meat." It is not a huge revelation that semiotic ruptures can take place; in the 1960s, Jacques Derrida began a whole method of analysis—deconstruction—based on the free play of relations between signifiers and signifieds (and it is worth noting here Derrida's conceptualization of deconstruction stemmed from his analysis of Saussure's dyadic system, not Peirce's triadic model). However, PBMAs do provoke interesting lines of thought about the ethico-onto-epistemology of meat.

While artificial intelligence continues to challenge human understanding(s) of what intelligence is, PBMAs challenge thinking about meat—the materiality, physicality of bodily configurations. Plant bodies can now enter complicated technological apparatuses and be configured to act—that is, smell, taste, feel, sound, look, and even cook (the recommended cooking time is that of beef patties)—like beef. What are the essential aspects of meat? In terms of achieving comparable modal disclosures (e.g., tasting a certain way, producing a particular smell), the animal body is no longer an essential aspect of meat.

Of course, the comparability of modal disclosures between animal and plant meat already depended on the animal body being the *absent referent* (Adams 2015) of meat products. PBMAs can alter dietary image vernaculars because processed meats already altered dietary image vernaculars. A beef does not function as a synecdoche for a deceased animal because the patty of a burger could not recognizably stand for the larger whole of a cow. The linguistic (e.g., pork for pig, beef for cow) and visual (e.g., frankfurters, hamburgers) separation of meat from animal body creates an opening in the dietary vernacular for meat to be plant meat.

I recognize that the discussion on PBMA disclosure has, thus far, stayed in the mostly theoretical realm of semiotics. However, with this understanding of how PBMAs disrupt meat semiotics, I am now able to discuss the ways in which these plant-based configurations afford new relational possibilities for people partaking in a plant-based diet, including new options for how one discloses a plant-based diet.

Feeling Closer

PBMAs create an ambiguous space where new attachments can form to plant-based diets. One post of meatloaf, mashed potatoes, and carrots was accompanied by text that stated that the person "veganizes" favorite meals when that person misses their mother and another companion and "feel[s] closer" to them (identity of account withheld). The photograph not only makes use of the *nema razlike* aesthetics of PBMAs to achieve closeness through likeness but also creates a sense of intimacy for viewers through the close-up shot. The food is shown from above, and the shot is zoomed in on the plate such that some of the meatloaf, some of the mashed potatoes, and parts of the carrots fall out of the shot. This is a tight, close shot, one that is so close that the field of vision is engulfed by the food but far enough that it allows each element of the composition to be seen at once.

The "feeling closer" photograph and text speaks to the forms of intimacy afforded by PBMA aesthetics and semiotics. This post indicates how the plant-based meat provides an opportunity to (re)create meals associated with a familial community while also maintaining individual commitments. At the same time, the *nema razlike* features of PBMAs allow the person to depict the meal through an intimate shot without it being obvious that this a vegan version of the family meal: the meal looks not like a "veggie meatloaf" but like a meatloaf.

For years, plant-based diets entailed a level of sacrifice that was not only about "the renunciation of real and irreplaceable pleasures" (Foer 2009, 78) but also about a profound change in one's sense of identity, often related to familial and cultural cuisine and traditions involving food. Across the hashtags, perhaps the most coherent function that PBMAs serve is enabling people to *feel closer* to familial and familiar practices. PBMAs showed up in pictures at fast-food restaurants, dine-in restaurants, pubs, and parks. They showed up in the dishes of many different ethnic cultural cuisines, as well as part of important subcultures (e.g., tailgating culture). Whereas vegetarian and vegan diets have historically been marginalizing and distancing aspects of one's identity, PBMAs are now included in and inclusive of numerous cultural cuisines and practices. Importantly for wellbeing, PBMAs allow people to participate in their ecological politics while also maintaining ties to familial and cultural identity, allowing flourishment in multiple domains.

This inclusion affords not only *feeling closer*, the introduction of PBMAs in grocery stores and eateries arrange space so people physically *are* closer. One person created a meme of their hand holding a bag of Beyond Jerky at a gas pump and the accompanying text was an exclamation of "hell yeah" to the fact that they could access vegan jerky at a gas station. Much has been written on the relationship between food sharing and human intimacy (Camposano 2019; Miller, Rozin, and Fiske 1998; Roth 1992) and on the stigmatizing feelings of *not* being able to share food with others (Bresnahan, Zhuang, and Zhu 2016). The discomfort of separation—physical and emotional—is an important consideration influencing when and how people choose to disclose their adoption of plant-based diets (Buttny and Kinefuchi 2020; Romo and Donovan-Kicken 2012).

The lack of aesthetic distinction between PBMAs and animal meats and the desire to *feel closer* brings me to a final point on how PBMAs disclose plant-based diets and identities differently, one closely tied to popularity and affordances of "food selfies" (Middha 2018)—or the sharing of digital food photography. Vernacular plant-based visual rhetorics, enabled by PBMAs and digital food photography, disclose—in both the personal sharing of one's choice of diet and the way in which the diet "shows up"—plant-based diets in a way that use the *nema razlike* qualities of PBMA images to background the dietary distinction and foreground the embodied desire of particular foods. Veganism, as a label, brings up negative connotations for many non-vegans. When the label is applied to people, it is often associated with a moral sanctimoniousness; when applied to diets, the label is associated with bland, unenjoyable foods. Social media platforms like Instagram afford the opportunity for people to *show* plant-based diets first, and to show them in ways that articulate with culturally desired foods: burgers are paired with fries and milkshakes, brats are smothered in onions between buns, plant-based chicken is shown with a sampler platter of dips, plant-based fish is wrapped in sushi nori and sticky rice, and plant-based ground meats are accompanied by pico de gallo and placed in taco shells.

Images of PBMAs correspond with familiar desires, and that is made evident through the common use of #foodporn in conjunction with each of the plant-based hashtags I followed. As of June 2022, there are nearly 290 million posts on Instagram cataloged with the hashtag "foodporn," eclipsing #vegan's 135 million, #plantbased's 40 million, #vegetarian's 35 million, #BlackVegan's 600,000, #BeyondMeat's 300,000, and #ImpossibleBurger's 130,000. The multi-cataloging capacity of hashtags—being able, for example, to tag the same image as both "#foodporn" and "#vegan"—enables plant-based images to circulate in wider networks and encounter different publics; it allows veganism to be part of practices of culinary tourism (Ranteallo and Andilolo 2017). This multi-tagging capacity of social media platforms is an important aspect of network- and community-making that allows hashtagging to be rhetorical and a form of activism (Jackson, Bailey, and Welles 2020). Further, the multi-tagging capacity urges deeper consideration of the idea that hashtagging reduces the universe to an index (Peters 2015, 320) and raises the question *what kind of index*? Where PBMAs detach the indices of a burger from a beef referent, hashtagging expands the notion of another index—the tool to catalog information—as received from codex media.

Many codex indices are, like hashtags, subjective tags that tell readers where to find a particular concept (Duncan 2022). The modern codex index is taken for granted as a fairly straightforward list of ideas in the book, but the index has always been a form of worldbuilding—full of decisions about what to leave in and what to leave out, how to present the ideas (including how to create a hierarchy of the ideas by deciding which terms are organizing terms for subsets), and are sometimes the site of vicious critique (e.g., as in a nineteenth-century historical text in which one author used the index to refer to a colleague's work as failing on many counts) (Duncan 2022).

Hashtagging a picture of a Beyond Burger at a bar with "#foodporn" and "#vegan" is not just an act of *cataloging* the image under the terms "#foodporn" and "#vegan"; it is an act that discloses plant-based diets differently. It moves plant-based diets into new networks and enables new associations. This is partly afforded by the way the social media indices work. To illustrate: say I was looking in a giant pictorial book on food, and I checked the index for "foodporn." If I were not eating a plant-based diet, I might skip any of the pages that corresponded to the subhead of "vegan" or "vegetarian." Or, if foodporn did not have its own entry but was only listed as a subentry to types of cuisine, it might not even occur to me to look under "vegan" for "foodporn." However, on Instagram, when I do a search and scroll through the #foodporn feed or add #foodporn to what I follow, I encounter all manner of foods that people have tagged "foodporn," including plant-based dishes.

Being first encountered by an image of food that elicits desires and then learning through the text that the image was of a PBMA is something akin to a visual version of the taste test. Further, it is not purely visual. As Alexa Weik von Mossner (2017, 52–53) writes, the visual is always multimodal, activating the endocrine system and occasioning consequences throughout the body. The multimodality of images is made particularly salient for this chapter by studies showing how images of food activate the production of ghrelin, the so-called hunger hormone (Seal, Gavaravarapu, and Konapur 2022). This "visual" taste test works similarly to the rapture-rupture process described by Tema Milstein and Charlotte Kroløkke (2012): there's the embodied experience of being captivated by something and that experience has the potential to create personal and cultural ruptures. First, a person may encounter an image of a cheeseburger, and then may be surprised to find that the saliva in one's mouth was activated by a PBMA.

In addition to the foregrounding of food photography, there is another way that feeling closer through shared desires was accomplished in the Instagram posts I studied—through what was backgrounded. I came to the *vernacular plant-based visual rhetorics* with a deep awareness of vegan and vegetarian rhetorics from both scholarship and activist efforts. I expected the majority of #vegan and #BlackVegan, and to a lesser extent #vegetarian, to include brief statements about industrial animal agriculture and the plight of animals raised in such conditions or the healthfulness of veganism or an assertion about greenhouse gases and climate change. In short, I was expecting each image to include a short position statement on *why* a vegan dish was being consumed. The pairing of an image with a position statement was the case in a minority of posts I examined. Sometimes, causes—like animal welfare or environmental activism—were associated with the image through the use of other hashtags, like #crueltyfree and #veganfortheplanet (which meant that a single image might be #foodporn, #vegan, and #crueltyfree), but more often the posts focused on enjoyment of the meals: Not how *good* it was to be vegan but how good it *tastes* to be vegan.

Vernacular plant-based visual rhetorics thus disclose plant-based diets differently by *not* readily disclosing them as plant based. Vernacular plant-based visual rhetorics featuring PBMAs emphasize a familiar and shared dietary image vernacular; they allow people to participate in recognizable, communal practices—like eating hot dogs at a tailgate party—while also attending to their individual dietary needs. Because PBMAs allow plant-based diets to show up differently, people are able to document and share their plant-based dishes in ways that invite *feeling closer* by reducing the distance between a plant-based diet and an omnivorous diet. Rather than disclosure through a morally valenced utterance, people can show plant-based diets—and themselves—as *nema razlike!*

BEYOND MEAT?

On a deep red wooden paddle—think red chestnut—on an even darker, brown wooden table—perhaps dark walnut—sits a large burger, the kind a person would struggle to hold in one's hands. The flat, golden bun creating the bottom of this sandwich has a white cheese—Swiss?—with splotches of hot sauce melting over the side. The meat patty is barely visible underneath the bowing, orange triangle of another cheese—probably cheddar—and some crispy, red sheets that suggests the presence of bacon. Onion rings and barbecue are stacked on top, and the sandwich is completed with a well-domed, golden, sesame seed bun. The only non-hashtag text accompanying the image is a request for viewers to "guess the brand," and below the text, the image is networked via twenty-eight different hashtags. Only one hashtag is #impossibleburger, which might suggest that this is a plant-based burger. However, the other hashtags—like #beefburger, #baconcheeseburger, and #bigburger—invite reinterpretation of what is meant by "#impossibleburger." It seems that the hashtag is not cataloging and circulating an entry of an image of the plant-based patty but rather "impossible" here is used to characterize the size and configuration of this meal. The presence of over two dozen other hashtags aids in deciphering what #impossibleburger might be referencing here, but consider another image of a burger sitting atop fries in a paper tray with over two dozen widely contradictory hashtags—one hashtag refers to #angusbeef and another to #impossibleburger, one to #vegan and another to #friedchicken (which was not pictured).

Because digital indexing—as cataloging—is *not* arranged like an index, how things are disclosed can be confusing. The system of tags is not as readily sensible as a codex index, and there are multiple hashtag generator programs available on the Internet to help users get more views on their posts. Sometimes it is obvious when a generator is used because the generator (e.g., Best-Hashtags) will place #bhfyp (i.e., Best Hashtag for Your Post) at the end of the recommended hashtag list; if the list the generator supplied is just copied and pasted, #bhfyp will appear at the end of the hashtags. Other times it remains a mystery how this composition of hashtags came to be, let alone what purpose the assortment of tags might be.

I refer to such image-hashtag pairings as *hashtag jams*, borrowing the term from Mark Dery's (1993) *Culture Jamming* and Naomi Klein's (1999) *No Logo*. Culture jamming refers to the practice of subverting mass media messages, an intentional practice done to raise awareness of social and environmental inequities. I use the term "hashtag jams" rather than "hashtag jamming" because hashtag jams can occur through both intentional (e.g., an image of a beef burger that used #vegan as a part of a larger post mocking plant-based diets) and unintentional (e.g., copying and pasting a generator's outputs) means. The jam occurs whether the jam was a purposeful act (e.g., satirizing) or not (e.g., using frequently accessed hashtags to increase views). The hashtag jam is a jam because it provides a moment of *stasis*—a stopping point—in which a person scrolling through a feed is paused, interrupted, or confused by a post's presence within a hashtag.

Stasis, a term with a long history in the Western canon of rhetoric, refers to "a series of steps or questions to ask which [lead] to the heart of the matter" (Kennedy, quoted in Jasinski 2001, 528). Traditionally, the questions or points of stasis inquire as to the facticity of a phenomenon, the kind of thing that phenomenon is, and what the procedures are for engaging with or responding to the phenomenon. Points of stasis are ways of finding the main elements at play in a dispute.

When a post of a burger of undecipherable origins enters a plant-based feed, it provides a stopping point in the feed because it provokes questions about what kind of thing the food is. Is it an animal meat burger or a plant-based burger? Jams work because they intervene in people's sense of *what goes with what*, or what Kenneth Burke (1984, 75) referred to as "pieties."

The jam is not necessarily intrinsic to the post—an image that "interrupts" my feed may not interrupt another person's—but when a person does encounter a hashtag jam, it is a stopping point because *something seems not to belong for some reason*. The same ambiguity afforded by PBMAs that provide openings for omnivores to find plant-based meals desirable and non-sacrificial is also what allows plant-based networks to experience jams. The traditional associations of omnivorous and plant-based diets no longer work. A burger that looks, smells, tastes, feels, and sizzles like beef no longer inherently *goes with* a carnivorous dietary item. Animal meat is not only disclosed differently to people; plant-based diets are disclosed differently, and they are increasingly disclosed in remarkably indistinguishable ways except for the ingredient list on the package.

If the ingredient list is *the* distinguishing feature on the consumer end of the product, then in important ways there is no *beyond* meat. PBMAs intervene in meat semiotics because they *participate* in meat semiotics; meat is still the source material for *imitatio*. Sometimes, the *imitatio* gets disclosed in endearing and interpersonally affirming ways, such as allowing someone to *feel closer* to missing loved ones. Other times, the *imitatio* of meat reveals troubling aspects of meat culture that the *nema razlike* features of PBMAs perpetuate. As PBMAs offer new opportunities for plant-based dieting—characterized by being *nema razlike* and *feeling closer*—it is important not only to celebrate the opportunity to participate in and be close to meat-eating cultures but also to ask with which aspects of meat-eating culture plant-based dieters want to be affiliated and which aspects of meat-eating culture are worth maintaining a certain distance from so that other relations may come into being (Giraud 2019).

For example, there's the "Beyond Burger Costume," a tight-fitting tube dress with the accompanying text that proclaims customers can "be the yummiest" patty at the party—complete with a "Not Grade A" branding symbol that will land on the left glute (to see the costume for yourself—as long as it is available—visit Yandy.com and search for "Beyond Burger Costume"). The site where the costume can be bought provides multiple shots of the model in the costume. In each image, one of the model's hips is angled to the side, emphasizing the model's curves. The page includes a "complete the look" heading with links to buy fishnet pantyhose that are advertised on a model who is only seen from the bellybutton down and a pack of lace, thong panties color-coded for each day of the week. The pairing of the model with sexualized food is characteristic of meat culture, only this dress comes with a headband displaying a tag that says "plant based" to alert viewers that the dress is a PBMA and not animal meat. What happens if the headband comes off?

Recall that Adams (2015) argued that hegemonic meat-eating culture oppresses both animals and women. Women were both criticized for plant-based diets that are considered "feminine" and make people weak, and women—as living, breathing individuals—were the absent referent in advertisements that marketed meat alongside models wearing bikinis. The models were filmed in ways to satisfy what Laura Mulvey (1989) called the *male gaze*, or perhaps a *toxic masculine gaze*. Both the anxieties and fetishes of "feminine diets" are part of PBMA discourse that circulates on social media. In addition to the sexualization of PBMAs discussed above, PBMAs are also depicted in association with emasculating anxieties. For example, there is the Impossible Whopper Estrogen meme that uses pictures of large-breasted women to satirize the claim by Dr. James Strangle that the Impossible Whopper contained eighteen million times the estrogen in a beef whopper than a PBMA and could lead to the development of breasts and gender transition. Often, the large-breasted women, sometimes real women and sometimes fictional animations, are also depicted in highly sexualized clothing and poses.

The issue here is not whether PBMAs are *good* or *bad*—certainly any entity can enter into relations that are either,[6] and there remains much debate on the overall healthiness and environmental consequences of the foods (Hu, Otis, and McCarthy 2019)—but is rather a matter of recognition that PBMAs contribute to both the flourishment and vulnerability of plant-based rhetorics, including vernacular plant-based visual rhetorics. PBMAs afford closeness to aesthetics and cultures otherwise closed off to plant-based dieters, and that closeness and lack of distinction is something that warrants ongoing scrutiny into what kinds of worldbuilding projects PBMAs are participating.

#BlackVegan offers some insight into how vernacular plant-based visual rhetorics can be distinctive in productive ways. Black veganism is an approach to veganism that attends to intersectional issues of health disparities for BIPOC communities, food inequity, social justice in unfair labor practices of industrial agriculture, environmental soundness, and the conditions of animals kept in confined feeding operation (Adewale 2021; Harper 2020). A critique of mainstream—or white—veganism has been that it can be an unsuitably simple answer to the complexity of *eating well* (Derrida 1991). A person may look at a label, see a green "V," and be done thinking about the food without ever considering its production, its nutrition, its accessibility. For example, quinoa is a popular food in many plant-based diets; it is rich in protein, fiber, and Vitamin B, and, unlike PBMAs, it is not a heavily processed food. However, the global gastronomic-economic boom that quinoa has experienced is a threat to the food sovereignty of Indigenous communities in the Andean region of South America (García 2013). The "V" does not disclose these relations. Similarly, the V on some vegan butters, also, conceals a host of harmful (more-than-)human relations—akin to what Garrett Broad (2016) calls "the social production of ignorance" (43)—such as the use of palm oil, which is often sourced in ways that decimate orangutans' ways-of-life and lives.

It is a challenge to eat well in a global, industrial agricultural system, and #BlackVegan posts do not meet all of the challenges and complexities of eating well, but the posts do tend to make foods recognizable, visible (e.g., it is difficult to decipher the foods that constitute a Pop-Tart). For example, quinoa was a common ingredient in the photographed meals, as in one meal that showed—and labeled in the accompanying text—quinoa pilaf served with black beans, sliced avocado, and pickled red cabbage. The foods were visibly distinct. Quinoa was visually decipherable as quinoa to people familiar with the food.

Further, #BlackVegan images were often accompanied by a kind of *citational politics* in the text. In addition to showing the food, recipes or at least a list of ingredients were shared. Instead of a caption naming the people found in the photograph, the captions worked to name the foods. The plants were not absent; they were shown and named. The citational politics of showing and naming creates a practice of knowability and accountability. Showing and naming foods work against the example of the unknowability of a burger simply marked #ImpossibleBurger. Of course, citations can be fabrications, but the act of naming one's food(s) and sharing even pared down recipes (e.g., lacking measurements, cooking times) participates in the nuanced food practices that Lewis (2020) claims work *against* dataveillance. Food citations are not going to "save the planet," nor are they going to monkeywrench dataveillance. However, the actions afford a measure of mindful engagement to digital food photography. Food citations *can* serve as a mechanism of care—a commitment to attribution and recognition—that helps people attend to the work of eating well.

A READINESS TO EAT WELL

The act of showing and naming plants is not alone an ethical practice, but it can be an important start for vernacular plant-based visual rhetorics. Showing and naming—identifying—is a practice that mediates perception and "can be seen as the basic entry to socially discerning" aspects of the more-than-human world (Milstein 2011, 4). In naming, one generates "certain kinds of ecocultural knowledge that constitute aspects of nature as considered, unique" (4). Showing and naming also makes one's inclusions and exclusions—what's on the plate, what's not—clear and is therefore an act of taking responsibility. As Eva Haifa Giraud writes (2019), there is no getting outside of entanglement—certainly not in the complex, global food web—but there are ways of making entanglements more visible, and therefore contestable and subject to revision.

PBMAs are now important actors in vernacular plant-based visual rhetorics, and PBMAs allow these rhetorics to surface in surprising places. PBMAs afford vernacular plant-based visual rhetorics the opportunity to come alongside meat-eating rhetorics—for consumers *feel closer* to customs and to loved ones and to eat according to one's preference in *nema razlike* ways. In other words, PBMAs are agents of/for ecological readiness: enabling more people to participate in plant-based dieting while living a whole life that includes participation to familial and cultural practices typically associated with the consumption of meat. PBMAs intervene in meat semiotics and change dietary (image) vernaculars because the *nema razlike* qualities create a space of ambiguity that is both exploited by PBMAs and is exploitable, as the hashtag jams make clear.

Vernacular plant-based visual rhetorics are flourishing in part due to the prevalence of PBMAs, and they are also vulnerable to becoming indistinct from meat-eating culture by the access that PBMAs permit. PBMAs can be an important part of empowering plant-based rhetorics and interpersonal relations, as long as the plant-based visual rhetorics that the PBMAs afford do not themselves become completely *nema razlike*—that is, as long PBMAs visual rhetorics do not lose all distinction from meat rhetorics. The interference of PBMAs in meat semiotics that allows PBMAs to problematize dietary (image) vernaculars—to disrupt the idea that *meat* is necessarily *animal meat*—also renders PBMAs vulnerable to co-option and circulation in hegemonic meat culture. For vernacular plant-based visual rhetorics to (1) maintain *plant-based* as part of their identity rather than just becoming *vernacular food visual rhetorics*, and (2) actually change the dietary (image) vernacular rather than being assimilated into it, there remains an exigence to maintain a commitment to *disclose* plant-basedness by showing and/or naming plants—in other words, making plants a present referent.

PBMAs draw attention to the ongoing process of disclosure and to the creation and problemization of vernaculars. It is no small feat that PBMAs can challenge taken-for-granted assumptions of what counts as *meat* and what it means to participate in dietary cultural practices.

The aesthetics of PBMAs allow the products to flourish as disrupters of dietary image vernaculars, problematizing the ability to know that a given *meat* is *animal meat*. At the same time, PBMAs' participation in meat semiotics also makes the radical aspect of plant-based diets—the abstention from dietary participation in the animal-industrial complex—vulnerable; it becomes harder to *see* plant-based diets in the world, even as they increase in abundance. How plant-based diets show up and act in the world is a relational, contextual affair, and the consequences of PBMAs over time depends on what ecocultural networks they get taken up by and help shape. For example, how do friends, family members, and colleagues attend to the vernacular disclosures of plant-based identities, and how and to what extent do these digital disclosures impact social gatherings? How do people attend to PBMAs as part of a nutritional diet and to what extent are people who eat PBMAs aware of the farm-to-factory-to-table processes and networks that afford a plant-based burger? These are important questions to be explored, but for now, it is sufficient to say that PBMAs allow plant-based diet(er)s to *show up* differently and there is a lingering question of how and to what extent are PBMAs *transforming* meat culture, and how is it transforming plant-based culture?

STORY TIME: (UN)BECOMING A PREDATOR

The lines between predator and prey seem straightforward. Nonhuman animals are often discussed as though they belong to one category or the other. When I worked for an animal welfare organization, I participated in a training on "large animal care" (e.g., horses, cows, llamas, alpacas) and a lot of the various species' behaviors were associated with them being *prey animals*. Similarly, when I took a training on working with birds, much of the raptor training emphasized their roles as *predators*.

Often, how one critter arrives in one category rather than the other seems pretty straightforward. Cats, raptors, snakes, sharks, crocodilians, many wild canine species, and dolphins are all obligate carnivores, creatures for whom 70 percent or more of their diet comes from animal meat. That puts them in the predator category. Horses, cows, llamas, alpacas, goats, and sheep are all herbivores; they eat plants, and they can be eaten by carnivores. Of course, my goats could absolutely be said to engage in predation (they consume sapling oaks), but goats are understood to be prey animals.

Then, there are chickens. Chickens, seemingly, are not just prey but *prey* animals, something close to the platonic ideal of what a prey creature is. I have lost companion chickens to hawks, foxes, and raccoons. I have accompanied my flock as they traverse the yard to go up for the evening, wary of the presence of a barred owl. Once while friends were over, I carefully removed an adult opossum from the coop. When I saw blood on the opossum, I initially assumed it was one of my chickens' blood, but actually the blood was the opossum's.

The blood was the opossum's. I thought I was protecting my chickens, but it seems the closed coop was protecting the opossum from the chickens. And it is not just the opossum. When my partner and I release the mice who wander into our home, we do so in a field that is away from our chickens. The ethics of domestication are troublesome. Because I brought my chickens to this space to live, I feel a responsibility for their wellbeing. Caring for their wellbeing includes providing shelter, food, water, and access to space to live a chicken life—to take dust baths and scratch up ground looking for insects and edible plant material. At the same time, I introduced the chickens into a forest ecosystem that did not have a say in their presence, either. The mice who live here have to negotiate North Carolinian wildlife, but the backyard chickens were thrust on their ways-of-life. It is a near-daily activity to ask some sort of question along the lines of *to whom am I called to care for in this circumstance?* Which leads me to my hardest-to-explain injury.

And then, there is me. The hardest-to-explain injury I ever sustained was a puncture wound from my generally gentle Buckeye rooster. I was chasing a hen who had a baby rat snake in her mouth. I saw the snake was still wiggling on either side of her beak, and I wanted to release the baby. My chickens are well fed, and it just seemed *bad* that this young snake should meet its end as a food source for a well-resourced hen. It felt like food injustice. Seeing me chase the hen, the rooster began chasing me, and when he caught up, he spurred the back of my leg.

I did not fault the rooster. The hen was free-ranging—exactly the life my partner and I wanted for her—and I was engaging in predatory behavior, chasing her with arms ready to swoop her up to remove the snake. Chasing and arm-stretched swooping are unwelcome behaviors in chickens' ways-of-life. These are predatory behaviors that often end poorly for the chicken. As I was trying to save this young snake, I was not *showing up* to my chickens as the primate-who-handfeeds-them-peanuts but as a primate-who-might-harm-them. I was, in that moment, engaging in predatory behaviors, behaviors with a relatively benign (taking food from another critter is hardly innocent) intent known only to me. As the rooster's spur punctured my skin, I was struck by how much *predator* and *prey* are transitory relations that get disclosed through actions.

NOTES

1. Pests are to animals what weeds are to plants: grouping categories to name beings living in places undesirable to humans (Brookshire 2022, 275).

2. A quick note on terminology: I use the term *plant-based* in *vernacular plant-based visual rhetorics* to indicate that these images concern the presence of plants and the absence of meats. *Plant-based* is a broad enough term to capture the many ways that one might approach a plant-based diet, like vegetarianism or veganism.

3. By discussing *consequences,* I am being careful to distinguish the plurality and ever-unfolding ways that these digital discourses and rhetorics act on and with bodies from *media effects*. *Media effects* refer to an area of social science study that explores how individuals and society are influenced by media.

4. In 1986, human rights activist Dick Gregory spoke at Amherst College about "the *plate* of Black America—about the politics, economics, health, and culture of what we ate, and why we should become vegans" (McQuirter 2021, 170).

5. Although I am discussing the signifying aspect of smell, the smell of burgers has asignifying aspects, as well. The smell of burgers activate pre-signifying affects of the sensation of smell and experiencing one's mouth watering prior to signification (cf. Muckelbauer [2021] on *asignification* as the compelling force of rhetoric).

6. The statement "certainly any entity can enter into relations that are either" is a recognition that there is not an ontological position from which PBMAs can be pronounced *good* or *bad*. Rather, they are situated in sociopolitical contexts that afford good/bad relations.

Chapter 5

Transformation

When the more-than-human world is disclosed differently, it affords opportunities for sustained attention to what was once backgrounded or taken for granted and for new affiliations to emerge. Transformation—the action of going beyond or across a creation or production—is possible. Ideas can change, values can change, practices can change. As shown in capter 4, people can come to see their diets differently and eat differently. Alternative ways of worlding become possible. For alternative worlds to take shape, transformation of our present sociomaterial conditions must occur. We are not engaging in worldbuilding from scratch but rather always in the process of generating *bricolage*—the construction and arrangement of new creations from available materials. Within the community-care paradigm, digital rhetorics of worldbuilding are aimed at transforming existing more-than-human relations in ways that afford the greatest power of living. As discussed in chapter 4, some people have taken to plant-based diets as a way of transforming consumptive practices within a world that nonetheless requires the consumption of other biological organisms for energy. Plant-based diets are a transformation of the practice of eating.

Digital worldbuilding rhetorics can also occasion transformation in how humans live with other animal species. Globally, humans can go from being largely unaware of the Australian marsupial known as the *quokka* to creating a social media #quokkaselfie craze (Holland 2015) that both transforms local tourism (with better and worse consequences) and dramatically increases donations to quokka conservation efforts (Bergman et al. 2022). The velocity and degree of transformation depends on a number of factors, including the possibilities afforded by platforms, the kinds of communication enacted by participants on platforms, and the natureculture context—that is, the biological, cultural, and personal factors that form the predispositional matrix (or ever-evolving subject positions) from which interactions occur—of the relationship in question.

Transforming human-quokka relations is different from transforming human-snake relations. Quokkas were largely unknown outside of Australia, but once images began circulating on social media, the "Muppet-cute" creatures were "irresistible" (Holland 2015). Transforming human-snake relations, however, meets with deeply entrenched biological and cultural factors that have contributed to a strong aversion to snakes for many people. This certainly does not mean that an *individual* human could not find a human-quokka relationship more challenging than a human-snake relationship. Rather, it means that owing to both biological and cultural factors, as well as personal histories and experiences, different interspecies relations *tend* to take certain shapes. In this chapter, I use rhetorical field methods of digital ethnography (that is, the joining of ethnographic methods like participant observation and interviews to the rich and well-established *grammar of rhetoric* [Ehninger 1968])—including participant observation and semi-structured interviews—to explore how human-snake relationship *tendencies* can be transformed through social media.

To do this, I first discuss the natureculture factors that contribute to the form that human-snake relations tend to take. Next, I describe how snake identification groups on social media work and how they provide a story to illustrate the transformative potentialities for human-snake relationships. Finally, I spend the bulk of this chapter analyzing how these snake identification groups use the process of identifying snakes to facilitate a transformational experience of identifying *with* snakes in invitational social media communities—in other words, how digital communities can invite transformation.

THE NATURECULTURE OF HUMAN-SNAKE RELATIONSHIPS

Humans have a long and complicated relationship with snakes, with feelings ranging from love and awe to hatred and disgust. For many people, *the only good snake is a dead snake*. Polls indicate that nearly two-thirds of Americans fear snakes, with half of those classifying themselves as "very afraid" (Moore 2014). The prevalence and intensity of ophidiophobia has made the fear a rich source of research for neuroscientists, comparative and developmental psychologists, and anthropologists, among others. Like all phobias, ophidiophobia can be debilitating or, at least, extremely uncomfortable. The sight of a stick can cause a heart to spin "out of order," as author Melanie Challenger (2018) described. While worldwide, about twenty thousand people die each year from an envenomated snake bite (Kasturiratne et al. 2008), this is the not the case in the United States. In the United States, there are about five deaths annually due to the bite of a venomous snake. There are fewer venomous snakes and greater access to life-saving antivenom in the United States than in many places, making death from snake bite an extremely rare occurrence.

It is rare to encounter a venomous snake at all in Emily Dickinson's (2022) home state of Massachusetts, but the poet echoed the feelings of many when she wrote that the sight of a "narrow fellow in the grass" made her breath tighten and chilled her to the bone.

Fear is also bad news for snakes. In the southeastern United States, people often kill nonvenomous water snakes thinking they are cottonmouths. Yet, "kill first, ask questions later" is a common strategy for snake identification (Jackson and Mirick 2022). Beyond the individual snakes' lives, human fear contributes to an apathy about the overall plight of snakes. For example, a team of biologists found that fear of snakes was associated with negative perceptions of snakes and the view that snake conservation is unimportant (Daltry et al. 2001).

One of the central questions concerning the fear of snakes is the age old one of "nature or nurture?" Are humans innately afraid of snakes, or do humans learn to be afraid of snakes? Snake phobia can be so intense that a picture of a snake can induce fear and panic (Marcum 2007). In that moment, the frightened person—desperate to get away from the frightening object—is unlikely to care whether the fear was passed down from an ancestor or learned from a frightened care provider. But *how* this question is asked still matters. *Natureculture* is a term introduced by feminist science studies scholar Donna Haraway (1998) to communicate the idea that nature and culture are so deeply entangled that they cannot be separated into "nature" and "culture." As a concept, natureculture challenges the ontological divide between nature/culture and human/nonhuman and gestures toward the deep relationality between sociocultural and ecological aspects of animals' lives. When the question becomes one of what and how naturalcultural factors contribute to the great prevalence of ophidophobia, the answers start to move beyond "just-so" stories (i.e., that circumstances just *are* a certain way), and beyond "just-so" stories lies the complexity of phenomena and possibility for transformation.

Stephen Jay Gould (1978) suggested that Darwinian evolutionary theorizing is a kind of "just-so" story. However, it can be productive to consider the long and rich evolutionary history humans and snakes share—not to paint a picture of a predetermined future or ontologic-closedness, but to consider how the human-snake relationship may have gotten to the place it is. Some scholars theorize that humans have an innate fear of snake, and that it is only through repeated, nonharmful exposure to snakes that people learn *not* to fear snakes.

There is good reason to think humans are innately afraid of snakes. It has been hundreds of millions of years since humans shared a common ancestor with snakes, which means there has been a very long history of a kind of biological arms race. Though both humans and snakes evolved from tetrapods, snakes not only shed their skin in a unique fashion, but they also came to shed their limbs (Lecointre and Le Guyader 2006). Imagine an early primate ancestor in a forest. This primate is adept at climbing trees, which gets it away from predators on the ground or in the water, but then there are snakes: a primary predator that can go anywhere the primate goes.

The primate's best defense was to quickly detect the snake and avoid it, necessitating capacities for pattern-finding and depth perception. In a classic Darwinian tale, the primates who avoided snakes reproduced, and over millions of years, the predatory pressure from snakes contributed to specialized visual systems in primates. Primates who evolved alongside snakes do have a disproportionately large pulvinar region in the brain, the region associated with the visual detection of relevant objects, and brain imaging studies have shown enhanced visual activity in the brain when humans are shown images of snakes. Humans come from a long line of primates trying to avoid becoming a snake's next caloric infusion. This is the thesis of Lynne Isbell's (2009) Snake Detection Theory, a well-supported idea that snakes played a significant role in the evolution of primates' visual systems. This evolutionary theory provides some insight into the *longue duree*—or deep history—of primate-snake relations. The theory also has profound rhetorical import in that it invites consideration of how species can participate in shaping not only the anatomical features (e.g., camouflage to hide) but also the perceptual world, or *umwelt* of other species (see Winthrop-Young 2010 on Jakob von Uexküll's theory for an extended engagement on the significance of different perceptual worlds).

Given this history, a *prepared association* of snakes with fear—the idea that humans have an evolved predisposition to associate snakes with fear—would seem a beneficial adaptation. However, a prepared association for fear does not mean that snakes are necessarily "innate fear stimuli in the sense that they automatically and invariably" activate fear in all individuals (Öhman and Mineka 2001, 487). To explore how these associations take shape, psychologists Vanessa LoBue and Karen Adolph (2019) revisit some studies of innate fears (e.g., snakes, heights, strangers, spiders), and start by asking *if* there is a fear present, not when and how it shows up. If one assumes that a fear is innate, any sign of avoidance might be interpreted as fear, but people regularly avoid things—like muddy puddles—without being afraid.

Lobue and Adolph (2019) suggest that humans have a perceptual bias for, rather than innate fear of, snakes. A perceptual bias is a state of heightened or preferential attention and is in alignment with what Isbell and others have found researching Snake Detection Theory. Primates are quick to take note of snakes. And not only snakes, but also long, curved objects (LoBue 2014) and patterns found on snake skin. There is a physiological response based on the rapid detection of snakes, but that response is not necessarily associated with fear (Isbell and Etting 2017).

Part of the difficulty for researchers is having to tease out things that—to the eye—look very similar, like startle and fear. A primate—human or other—may jump or freeze at the sight of a snake before consciously processing that they saw a snake. To an outside observer, this could look a lot like fear, as opposed to a startle response. Also, the startle response might make someone more likely to then label the feeling as fear. Startle may be associated with but should not be conflated with fear. The startle response is the preconscious-awareness moment of detection; fear requires assessment. I will return to this distinction later in the chapter. For now, it is important to register that being startled is not always associated with fear. Someone might be startled by a surprise party, but the startle response in that scenario typically gets articulated with joyful feelings. I need to elaborate on the more culturally valenced aspects of the human-snake *natureculture* relationship to help understand how the startle response became associated with fear.

Heidi Marcum (2007) describes the cultural complexity of humans' relationships with snakes. More affirmatively, snakes have served as symbols of fertility in ancient Crete and Mesopotamia, as a symbol of immortality (Oroboros), as key figures in the symbol of medicine (Caduceus), as gods by Hindus in India, and as a zodiac symbol representing wisdom, patience, and confidence in China. Less affirmatively, ancient Greek depictions of Medusa and Hydra portray snakes as villainous monsters, and Judeo-Christianity puts snakes in the middle of the Original Sin and humans' expulsion from the Garden of Eden.

Snakes are also present in the oral traditions of people indigenous to North America. There are stories of a "wicked woman" who had affairs with snakes or an "overly proud" woman who gave birth to the first snakes. Similar to the story of Original Sin, a Peoria story tells of a woman tricked by a snake—a rattlesnake who took the guise of a handsome man. An O'odham legend tells a story in which the rattlesnake was responsible for bringing death to the world. However, in this legend, the snake only brings death after seeking help from Elder Brother because other people and other creatures kept poking the rattlesnake to hear his rattle for their own enjoyment (Native Languages of the Americas 2020).

There are also many so-called old wives' tales surrounding snakes that paint a terrifying picture: Coachwhips will chase a person down and whip them to death with their tails; snakes cannot be killed unless their bodies are cut into seven pieces and burned in seven different fires. There is a vast, historical, and multicultural library of snake stories from which to choose, and a common thread in many is that snakes are agents of trickery and often of evil, creatures to be feared.

Depictions of snakes in contemporary popular culture often tap into and perpetuate fear of snakes, such as in popular movies like *Snakes on a Plane* or in the book/film series of Harry Potter. Fans of Indiana Jones may remember that snakes were a sort of kryptonite for the intellectual action hero. Lest one think this kind of negative representation is limited to action-fantasy genre, snakes even show up as "cold-hearted snakes" in popular music or used in news coverage to describe a serial killer as a "serpent in human form" (*Catching Killers* 2021). Media depictions of animal-to-human violence have been linked to high levels of fear and estimation of personal risk of injury, even when paired with public service announcements that attempt to educate people about the animal and explain its importance to its ecosystem (Myrick and Evans 2014).

There are also familial and communal aspects of the human aversion to snakes. Many people are brought up in households in which they hear stories about bad experiences with snakes or witness adults in their lives reacting adversely to the presence of snakes, and there are the ever-present toy snakes sold to prank and frighten (e.g., in some stores, toy snakes—along with toy spiders, roaches, plastic poop, and whoopie cushions—are all sold together under the banner of prank toys).

Susan Mineka and Michael Cook (1986) found that group effects of fear-modeling are also present in rhesus monkeys. Monkeys exposed to other monkeys who were fearful of snakes were more likely to behave fearfully, and monkeys who observed other monkeys behave non-fearfully around snakes were less likely to behave fearfully themselves. The authors observed that social modeling plays a significant role in how and whether fears become amplified or abated.

The biological differences between human and snake bodies combine with the familial, communal, cultural, and medial postures toward snakes to produce *rattlesnake roundups*. Many people grow up in communities in which snake bounties are part of the local history (Levin 2016), and every year, tens of thousands of rattlesnakes are still slaughtered in rattlesnake roundups in the southwestern United States. Proponents of the roundup cite concern for the safety of humans, pets, and livestock, but the activities surrounding the event are evocative more of fairs, fetes, and festivals than of public health and safety interventions.[1] Although existing in several states, Texas hosts the largest roundup, complete with the collection and subsequent extermination (through starvation, suffocation, or decapitation) of hundreds of thousands of rattlesnakes annually (Schipani 2018). One popular method of collecting snakes is a technique called "gassing," which involves pouring gasoline in dens to flush out the snakes. Gassing is detrimental to a number of animals other than snakes and is an environmentally harmful practice.

As Hal Herzog (2010) observed: when it comes to animals, there are some people love, some people hate, and some people eat. This natureculture sketch of human-snake relationships does not seem particularly promising for finding resources for inventing ways to live together otherwise. Thankfully, there is more to this story.

GETTING SOCIAL WITH SNAKES

Social media platforms are part of the human-snake interspecies relationship and even a brief search of "viral snake video" indicates that the content on many of these platforms show snakes as tricksters at best and monstrous at worst: "Snake Tries to Steal Fisherman's Catch," "Viral Video Shows Snake Eating Itself," "Watch the Viral Video of a Huge Snake Lifted by a Crane," "Snakes Battle in a Viral Video Shot in Australia," "Snake Tries to Choke Kerala Man," "Viral Video: King Cobra and Python Fight Aggressively," "Crazy Snake Bite!" and "Viral Video: Snake Tries to Follow Toddler Inside the House." The rhetoric of clickbait makes visible some important aspects of the human-snake relationship in these titles. While viral dog videos have titles likes "Viral Pet Videos That Melt Hearts" and "Dog Paddles Through Muddy Water to Save Deer" and evoke an affectionate interspecies bond, the snake videos that circulate have titles that tap into anxieties about the serpents' aggression ("battle," "aggressively"), size ("huge snake"), perceived malevolent intention ("steal," "choke," "crazy snake bite," "follow toddler"), and distant otherness ("eating itself").

A Brandwatch search for "snake" (with the filter for specific references pertaining to the reptile and not, say, the classic video game or the '80s hair band) from January 2021 to January 2022 across Facebook, Twitter, YouTube, Instagram, Tumblr, and Reddit report the top three emotions associated with snake content as fear, disgust, and neutral. As one might guess, the same search run for dogs comes back with a top emotion of joy, followed by neutrality and sadness. The fear and disgust people express do not always remain online. Some "field herping" sites and groups—digital media dedicated to sharing details and photos of people finding snakes—contribute to the deaths of snakes as some people use the location information to find and kill the snakes (Levin 2016).

While far from an exhaustive account of how snake content serpentines through these digital networks, this brief sketch provides a sense of the semiotics and emotional intensities associated with snake videos, images, news, and other posts, and is sufficient to register that there are substantial resources—from cultural narratives to biological theories to mass and social media—that support an adverse human-snake relationship and make mutual flourishment a challenging affair.

But there are other communities where snakes show up differently. Offline, there are science and education centers that promote the message "the only good snake is a live snake" (Rosenfeld 2021); online, there are *snake identification groups*. Researchers and science writers have taken note of these groups and the consequences of their existence; recently, an article in *Scientific American* boasted the headline, "How Facebook Is Saving Snakes" (Willingham 2022).

Snake identification groups refers to groups on social media whose organizing principle is the expeditious and accurate identification of snakes in images uploaded to the group. This work has also been referred to as crowdsourcing snake identification. In January 2021, there were approximately fifty snake identification groups on Facebook, some dedicated to worldwide identification, some US-state specific, some dedicated to identification and additional educational discussion.

Although I have spoken to people who identify snakes on social media platforms like NextDoor and Reddit, my participant observation took place in Facebook communities and all of the moderators and community members I interviewed were affiliated with snake identification groups on Facebook. The emphasis on Facebook is not intended to privilege Meta, Inc. or suggest that what's being done on Facebook is not or could not be done on other platforms. Facebook was selected for this particular community because: (1) Facebook is, at the time of this writing, the leading social media platform in terms of sheer number of users worldwide (Statista 2022); (2) members of snake identification groups with whom I spoke shared that Facebook snake identification groups were both recommended by friends and suggested by Facebook's algorithm; (3) identification on other platforms often seemed more spontaneous, such as a snake found in one neighborhood on NextDoor, as opposed to snake identification groups on Facebook whose group identity is constituted by the identification of snakes; and (4) the Facebook groups have community rules that maintain fidelity to group identity (e.g., posts are deleted that do not pertain to snake identification).

These snake identification groups are typically comprised of moderators and administrators (admins) who monitor the posts on the boards for compliance to community standards and who close off commenting to make sure that newer posts are seen and attended to promptly. The moderators and admins are also frequent contributors to identifying the snakes in the photos, as well as other group members who spend time on the board. People who identify snakes range from herpetologists and biologists to snake hobbyists to people whose primary engagement with snakes is through the mediation of snake identification groups (Durso et al. 2021). In these communities, snake identifications typically consist of a common (e.g., black rat snake) and scientific (e.g., Pantherophis obsoletus) name, and a note about whether the species is venomous or not. Some identifiers for venomous species include short remarks like, "best admired at a distance."

One major snake identification group on Facebook, called Snake Identification, is a public group that began in 2008 and now has nearly 250,000 members. This is a long social media life span considering that Facebook started in 2004 and did not open to people other than college students until 2006. Over the years, a number of similar groups have been created. Wild Snakes Education and Discussion is dedicated to longer-form discussions about snakes, whereas Snake Identification is meant to identify a photo of a snake quickly and accurately and then move on to the next snake. WSED is a private group that began in 2015 and now has nearly 85,000 members. While Snake Identification has an international board, many states in the United States have boards dedicated to making state-specific identifications, such as N.C. Snake Identification & Education (a group that began in 2018 and has over 47,000 members) and Arizona Snake Identification and Questions (a group that began in 2018 and has nearly 19,000 members).

In January 2021, Andrew Durso and colleagues conducted a study that attests to the accuracy and usefulness of crowdsourced snake identification. Several of their findings have rhetorical significance. One, the authors found that expert identifiers, who are both highly accurate and quick, make identifications through gestalt reactions rather than taxonomic keys; some reported having difficulty articulating how they knew, saying they "just knew" (14). Two, even though experts had high accuracy rates (between 81–95 percent accuracy depending on the level of identification, that is, species-level, genus-level, or family-level, with more accuracy at the higher taxonomic levels), communities tended to fare even better at making identifications. Three, people were better able to identify snakes local to their region. Durso and colleagues also spoke to the life-saving potential of such groups, for example, by increasing the capacity of health care workers to obtain snake identifications quickly, specifically, and accurately.

The potentially life-saving function of these group is something administrators of the boards are aware of and take seriously. The responsibility administrators feel toward quick and accurate identification is also made clear in these communities' rules and processes. Administrators, especially on boards designed solely to provide fast and reliable identification, closely monitor posts and shut down commenting to make sure the newest posts seeking identifications get a timely response. It is not solely the human lives the contributors to these boards are working to save, though; they also are hoping to transform how people understand and live with snakes.

Before describing how these rules and identifications help transform how people live with snakes, first, let me show *what* kinds of caring transformations can occur by sharing Erica's story.

Transforming Erica

A few years ago, on a sunny August afternoon, Erica (a pseudonym) took her eight-year-old son kayaking. Their orange vessel stood out against the murky creek water and the tall grasses and trees flanking them on either side. As they rounded a bend in the creek, something else stood out. "I could see him on the side [of the creek]. I didn't know what to do," Erica said. She had been told her whole life that snakes chase people, and now her kayak was about to cross the path of a venomous cottonmouth.

"I can't get in the water—the snake is there. I can't go anywhere. I can't run away, can't climb a tree. . . . I wanted to panic, but my son would have panicked, and I didn't want to scare him."

Erica stopped the kayak and grabbed her phone and a paddle. The paddle was for defense; the phone was to provide evidence in case anything happened to her and her son.

Erica's toes curled as the snake approached the side of the kayak. She gripped the paddle. It was all so quiet, except for the sound of her heart pounding in her ears. She tapped the paddle once against the side of the kayak.

The cottonmouth vanished under the boat. "Where did he go?" she whispered.

He emerged on the other side of the kayak and . . . "went on his merry way."

"It was opposite everything I've ever been told," Erica said.

Erica can recall a time as a child when her mother ran over a snake. The myths surrounding snakes were so exaggerated, their capacity to harm so magnified, that she recalled picking up her feet from the floorboard and screaming as the car went over the snake, fearing the snake would be able to get in the car and hurt them. Yet, there she and her son were, exposed in a kayak, and the cottonmouth seemingly took no note of them. Erica began to wonder what else she had wrong about snakes and if she could come to feel differently about them. She joined Wild Snakes: Education & Discussion, a social media group with over sixty-five thousand members, to learn more. The group is a diverse group of people, from those with snake phobias to those who have dedicated their lives to caring for snakes, who come together to discuss snakes.

"I just knew other people weren't afraid, so I didn't have to be either," said Erica.

After spending time on the board, Erica's attitudes, beliefs, and feelings about snakes began to change. She knows the difference between how being startled—that perceptual bias and immediate affective registrant—and being fearful feel in her body. She no longer considers herself afraid of snakes, but said that if she sees one, there is "that split second of surprise." "They [snakes] startle me," Erica said. "But," she continued, describing how startle differs from fear, "that's not the same with spiders. They make my hands sweat. Sitting here talking about spiders makes my hands sweat." However, once that split second—the rapid detection—is over with snakes, she is "able to contain" herself because she is no longer fearful of a snake chasing her. The threat assessment has changed.

About a year after we spoke, I was surprised when I saw a message from Erica in my Facebook Messenger. A friend of hers—who is not a member of Snake Identification—posted a picture of a snake in that person's yard and was concerned if the snake was venomous. Some friends immediately responded that the snake was venomous—a cottonmouth—but Erica made the right identification (a nonvenomous water snake) and explained how she triangulated the markers to make the identification. Her thorough and confident identification—knowledge she gained through prolonged engagement with Snake Identification—persuaded the friend that Erica's ID was right: this was not a life-threatening snake. Without my asking, Erica screenshot the exchange and sent it to me, saying she was incredibly proud of herself.

Erica's story underscores how "all relations compound without limit" (Deleuze 1988) and shows how her care for her son and a paradigm-disrupting encounter with a snake opened her to new possibilities of existence, of ways of living with snakes. After the encounter she joined a worldwide network of people on a social messaging service dedicated strictly to identifying snakes, and by spending time on that board she came to see differences in snakes, to see different kinds where once she just saw *snakes*—and to see *individuals*, and to attend to *behaviors* and not just venomous-nonvenomous groupings. She was then able to share this information with someone else—who is not in the Snake Identification network—who then decided *not* to kill a snake in her yard.

While encounters like Erica's are not in themselves sufficient to address the macroscopic problems of the Anthropocene—climate change, loss of biodiversity, habitat loss, pollution—they do help to cultivate new relations in the web of relations. Vibrations in the web are neither unidirectional nor immutable. Where Erica-encountering-a-snake used to be a diminishing experience—that is, it decreased the wellbeing of those involved—for many parties, now Erica can encounter a digital snake and provide information that spares a physical snake's life. That is a different vibration in the web of relations.

Now, it is time to discuss how these snake identification groups function rhetorically to transform human-snake relations.

TRANSFORMATION AT A DISTANCE

The snake identification groups are, clearly, dedicated to the work of identification, and this practice of identification works in both taxonomic and social dimensions. However, the identifications do not happen in a purely transactional, information-gathering environment, like one might expect from using a nature-identification app on one's phone. While these apps have their own rhetorical procedures (Bogost 2010), the snake identification boards are run by humans who craft a different rhetorical space—a rhetorical space that enables a sense of community. I will first go over how these boards' community norms invent an inviting digital space. Next, I will discuss the taxonomic practices of identification before attending to the social dimensions in which these practices occur, and finally, address how the interplay of these dimensions afford transformative human-snake relations for individuals-in-communities.

Crafting and Maintaining Community Norms

While Durso and colleagues' work (2021) focused on the accuracy of identifications and their contribution to medical treatment, I am interested here in how the rhetorical nature of identification within these digital groups works to foster new communities, ones that consist of human and more-than-human beings. What these online snake identification groups share is that they engage in consistent, predictable practices of identification—with the predictability achieved largely through community norms—to establish new possibilities for living alongside snakes.

Beyond the rapid and straightforward identifications, much more is going on. For example, administrators of these boards set up a "no shaming" rule (typically the first rule listed in community guidelines) and maintain adherence to this rule by issuing warnings for violations and banning repeat violators from the group. Community rules articulate what counts as appropriate, *decorous* behavior within the community and indicate what constitutes an act of shaming. For multiple snake identification communities, shaming includes the use of emoticons (e.g., using an "angry" response when someone posts a picture of a dead snake). Community decorum may also discourage posting pictures of a dead snake, but the rules expressly prohibit other community members from responding to such an image with an angry face. One group on Facebook has as its first rule that the group is dedicated to "education and understanding" and that no one learns through insults. Thus, even if a picture of a dead snake does come through, people should not respond with, "Too bad the snake's dead" or "Why did you kill it?," and just provide a simple identification of the snake.

Board members of snake identification groups reflected on several moderator techniques that made the board more than just a place to go for a quick, transactional exchange of information, such as the diligence of moderators to catch and correct any mistaken identification, the way the moderators quickly closed threads to make sure the next post in the queue was seen and responded to promptly, and the maintenance of community standards so that people felt safe asking questions they felt they could not ask elsewhere. Before getting into the rhetorical consequences of being able to identify snakes, it is worth spending a bit more time on the intersection between community standards and care because many participants shared that it was the maintenance of respectful community standards that provided the conditions for them to be engaged members of the group.

What community members described to me about their experience on the board was a sense of being welcomed and respected without expectation or pressure. In many ways, participants described a sense of *invitational rhetoric*. *Invitational rhetoric* is described as "an invitation to understanding as a means to create a relationship rooted in equality, immanent value, and self-determination" (Foss and Griffin 1995, 5).

Invitational rhetoric does not purport to be the *ontological* definition of rhetoric; rather, invitational rhetoric is a species of rhetoric in an array of rhetorical forces (Bone, Griffin, and Scholz 2008). In other words, invitational rhetoric can work alongside other worldbuilding aspects of rhetoric. Within invitational rhetoric, two primary rhetorical forms are assumed: (1) the offering of perspectives through personal narrative, nonverbal communication, and re-sourcement (i.e., responding to a message from a different framework or perspective); and (2) the creation of the external conditions of safety, value, and freedom. These external conditions, as well as others identified by Foss and Griffin (2020) in their initial position paper, were significant factors in members' engagement with the boards. More specifically, members shared how the boards communicated:

- *safety*, a sense of security rooted in the knowledge that agents will not attempt to degrade or belittle: "[board members are] very, very nonjudgmental"; "[the admins/moderators] do not condemn or belittle or let others do that"; and "there's no shaming [by anyone on the board]"
- *value*, the perception that people involved feel worthwhile: "somewhere in the answer is a supportive comment"; and when people express embarrassment over—for example—saying "poison" instead of "venom," a community member "will say, 'It's OK, you're learning'"
- *freedom of identity*, the feeling that identities are freely chosen: "I can follow or unfollow at any time"; "I'm in control"; and "I can always unjoin, and it [the board] could be helpful"
- *choice*, the feeling that multiple opportunities for personal development exist: the rules are clear because the admins/moderators understand that not everyone loves snakes; the board is "a growth opportunity"; a person can be a member without ever posting, can "gamify" identifications for one's self, can post photos, can contribute to identifications
- *autonomous interdependence*, or what I call attention to *individuals-in-communities,* the care and respect involved in the recognition that

individuals are simultaneously autonomous agents and interdependent with others: the boards offer a "collective community" with compassion for each member and for "the critters"; newcomers are welcome and gently corrected when they violate rules; and "the human component is everything. I will take a shot of happiness and community wherever I can find it"
- *order*, a means for individuals to engage in sensemaking: administrators are clear about rules and that guessing on an identification is not allowed; the "rules are clear and easy to follow" because one "knows what is expected"; and the format is "predictable": a picture will be posted and an identification will be offered

In invitational rhetoric, the means constitute the ends. In other words, invitational rhetoric is more aligned with ethics of care frameworks than consequentialist ethics. Despite the context-specificity of invitational rhetoric and ethics of care, an ethical principle animating both invitational rhetoric and ethics of care is the "recognition of the immanent value of all living beings" (Foss and Griffin 1995, 11).

When I first encountered the online snake identification and education communities in 2018, I was skeptical that they would restrain themselves and others from condemning acts of violence toward snakes. Perhaps it is more accurate to say that I was skeptical that it was a good idea to avoid condemnation. In line with critics of invitational rhetoric (cf. Condit [1997] and Jørgensen [2007], who offer rebuttals to Foss and Griffin on the importance of persuasion) and from my background in humane animal advocacy, I understood an outright attempt to persuade people to see snakes differently was not a violation but a constructive and ethical act of human communication toward the goal of multispecies flourishment. One could imagine an analogous dog breed identification group in which people posted images of dogs they had killed with a comment like, "This dog was in my yard. What breed was it?" It is unlikely that such posts would not result in legal involvement, let alone be tolerated and even embraced as part of the "learning process."

However suitable this comparison may be from a deep ecology perspective of the intrinsic worth of individual lives, it misses important cultural distinctions and reduces the complexity distinguishing human-snake relations and human-dog relations. This caution around comparisons is perhaps best explained by Claire Jean Kim (2018) who argues against the comparisons made in (more-than-)human suffering, particularly between animals in the industrial food complex and humans kept in slavery. When such comparisons are made, the unique horrors of both are obscured. Kim argues that it is possible to attend to logics that afford human and nonhuman animal suffering, such as a colonial mentality of hierarchy and separation, without losing sight of *how the enactments of those logics* and the consequences of those enactments are distinct.

While snakes have an ontologically equal claim to their lives and flourishment as dogs, it is not helpful to pretend that the sociocultural context of human-snake relations is the same as that of human-dog relations. While the ancestors of humans and modern snakes have long engaged in a biological arms race, with the scales being tipped decidedly in favor of the humans in only the last couple of centuries, dogs and humans are thought to have profoundly domesticated one another (Haraway 2003). Thus, what seems at first as a too permissive attitude in snake communities may actually be a powerful rhetorical strategy for building community, one that acknowledges the complex naturecultural evolution of human-snake relations. As John Elder (2014) writes, it is "[p]recisely because the ecological challenges are now so urgent . . . we require a more invitational, inclusive, and personally engaged approach to environmental activism" (xi). The community rules and the ongoing work of moderators to maintain them create the conditions for an inviting space to learn to identify, and potentially identify with, snakes.

Taxonomic Identification

The first and most obvious aspect of the snake identification groups is the practice of identifying and teaching others to identify snakes. Photos are posted to a group, and within minutes, one or more people will offer an identification. Usually, these identifications include both common and scientific names. Often, there will be some brief comment accompanying the identification; for example, whether or not a snake is venomous. In some groups, an identification will be accompanied with a description of the identifying features. Speaking to members of two different states' snake identification boards, the presence or absence of this last feature—sharing the identifying features or not—was associated with community members' overall satisfaction with the community. In the state group that did not share identifying information, members expressed dissatisfaction at being left to guess what distinguished one snake from another. After all, one member said, his goal was to learn to be able to identify for himself. In the state group that did share identifying information, members expressed feeling empowered by the information they were learning.

Although groups work to identify a species, the emphasis of the identification work is on whether a particular snake is venomous or nonvenomous. The venomous/nonvenomous designation is often what people who encounter snakes in their yards or who have been bitten want to know. It is worth pausing a moment for some rhetorical insight into the practice of identification and binary classification (i.e., venomous or nonvenomous).

At first glance, this dichotomous classification would seem to be problematic. Binaries tend not only to separate but to create hierarchies. In this case, the nonvenomous/venomous binary would favor nonvenomous because—from an anthropocentric perspective—nonvenomous snakes do not pose the threat to human safety that venomous snakes do. The nonvenomous/venomous division could function to synecdochally essentialize all snakes into two groups through the presence or absence of venom. Of course, this binary—like all binaries—does not *quite* work. In eastern North America, for example, a prevalent species of venomous snake, the copperhead (*Agkistrodon contortrix*), has an envenomated bite that is readily treatable, even for children, dogs, and cats (Wills et al. 2020). However, in Asia (e.g., reticulated pythons), South America (e.g., anacondas), and Africa (e.g., African rock pythons), there are large constrictors who are capable of killing adult humans. Further, animal educators try to get people to view the presence of venom *from an ophidian perspective*. From this perspective, people are invited to imagine having to hunt for food. In this fantasy, a person would either be equipped with venom and be able to strike their prey once, retreat, and wait for the prey to die to then consume it *or* would need to wrap around the prey and constrict it until the prey died, risking harmful bites and scratches while this happens (Rosenfeld 2021, 11).

So, it would be concerning if the labels of "venomous" and "nonvenomous" were performing the bulk of the work of identification. One could imagine a board in which *all* someone did was respond to a photo by saying "venomous" or "nonvenomous" without offering names. And, as already stated, the ability to discern venomous from nonvenomous *is* a primary concern of visitors to snake identification communities. Many members I spoke with shared that they are absolutely confident in their ability to differentiate a venomous snake from a nonvenomous snake but not as confident in anything more specific than that or that their strategy for learning snakes was to focus on learning the venomous snakes first. However, something interesting happens when people learn to distinguish venomous and nonvenomous snakes: "The invisibility fades away," one community member told me, highlighting that knowing something—anything—about snakes or a snake makes it visible, salient, and particular.

The member elaborated that, "If I think of snakes as one category, then they're all dangerous." However, in learning to identify snakes, in becoming confident of one's ability to distinguish features and thus make the encounter less ambiguous, even venomous snakes become less scary. This particular community member described how he and his spouse moved to their dream home and the *only* drawback was that their dream home would be in an area with snakes. He joined a snake identification group to learn which snakes he might encounter and which of them would be dangerous. He told me that he still does not want to touch a snake, and he would not mind if he never saw another snake, again. Yet, one day out on a walk with his spouse, they came across a timber rattlesnake (*Crotalus horridus*). Instead of running from the snake, he and his spouse stopped and captured a short video of the snake crossing the path. He was anxious, and "it was cool." *The invisibility had faded away.*

What is striking—no serpentine pun intended—about the comment that the *invisibility fades away* is that it starts from the negative; the default state is that snakes are invisible. This is perhaps partly a physical phenomena: snakes seem to "vanish on the forest floor," as one participant acknowledged. Their capacity for camouflage helps explain why it was so important for primates to point out their presence to one another. It also speaks to what ecolinguistic scholar Arran Stibbe (2021) calls the *erasure* of animals from human life (141). Animals are made invisible in human texts and practices through absence (being completely omitted), minimalization (being backgrounded), and/or masking (being present but in a distorted form—whether it be making the snakes appear monstrous or "Bambifying" them with large, human-like eyes to make them more appealing). Snakes are often invisible through each of these mechanisms.

Part of what the taxonomic work of identification does, then, is to make snakes *visible*. That is, the snake identification boards help to make snakes just *show up* in people's lives. Rhetorically, visibility functions to (1) evoke commonality through moments of embodiment, (2) establish the recognition of others through particularity, and (3) challenge taken-for-granted ideas (Gallagher and Zagacki 2007, 113). In learning to identify—that is, to name individual species of snake where once one just saw undifferentiated *snakes*—community members find snakes are made more visible to them. Community members shared with me not only their own in-person encounters with snakes since joining snake identification groups but also how pictures of others' interactions with snakes surprised or impacted them. One person described being taken aback by a photo that a woman posted. She had stepped on a snake in her garage, and she was not bitten but still wanted to know on whom she had stepped. The answer was a copperhead. The community member described this as a particularly impactful moment, as this was about as far from a snake being "out to get someone" as could be imagined. This was a common experience for community members with whom I spoke. One person vividly described to me her reaction to snakes prior to joining a snake identification group: If she were outside and someone shouted "snake," she would run inside, lock the door, draw the curtains, and be waiting with a weapon to defend herself. After seeing image after image of snakes not displaying any interest in the human photographers, she began to question how much snakes really cared about humans. Now, she no longer gets heart palpitations when she sees snakes, and if she were in the house when someone shouted "snake," she would go outside to see. This, she said, is "liberating." Now, snakes not only show up, but they show up *differently*.

The empathy afforded through the recognition of particularity is, perhaps, the most surprising aspect of how taxonomic identification renders snakes visible. Unprompted, almost every community member I spoke with expressed their sadness when a picture of a dead snake shows up to be identified. One member described to me the worry she feels when there is a post with an ambiguous outcome from this human-snake encounter. She will watch the board's feed and say, "I hope these people didn't kill that snake." She elaborated, saying that she finds it remarkable that she is now "worried about some snake six counties over." Another member described how the pictures of dead snakes "hurt my heart." Multiple people with whom I spoke had either killed or been with someone who killed a snake in the past; now—as one member described—it's not just a passive "live and let live" attitude but an active "I will go out of my way to help a snake." From stopping a car to let a snake cross the road safely to intervening when someone intends to kill a snake, people expressed the pride they took in their new relations with snakes. One person even labeled it a redemption story.

Being able to see the particularity of snakes is a process of going from seeing snakes as a monolithic group of limbless, elongated, scaly herptiles to seeing "clusters of traits," as one participant stated. The capacity to recognize distinctions affords a practice of care—taking care to notice and attend—into the act of scrolling through one's feed. A person's capacity for recognizing distinction allows images to hail them *differently.* Imagine how long it would take to complete an image-based CAPTCHA that asked participants to choose all the images containing snakes. Maybe some images would contain ropes or worms to make it more challenging. Now, imagine an image-based CAPTCHA with only pictures of snakes, and participants are asked to choose the images of corn snakes. The capacity to even complete such a task requires, as one participant described, "learning to see snakes" like the moderators do.

Let me unpack the significance of seeing snakes as the moderators do. In Durso and colleagues' study, the researchers found that expert identifiers did not typically use identification tools. Rather, they identified "by gestalt"; they "just knew" (Durso et al. 2021, 14). Gestalt is the cohering of disparate parts into a whole; they are unconscious but absolutely sensorial (Poole 2020). It is logical, then, that identifiers would find it hard to put into words how they knew. What is interesting about this response, though, is what it reveals about the orientation toward snakes by people who identify by gestalt.

Social scientists have long known people make more distinctions among in-groups—see more variation in the appearance of in-group members, expect more differences in behaviors, and so on. The opposite of this—the expectation of homogeneity among out-group members—is how stereotypes get formed. When people recognize someone with whom they are close, they do not perform a checklist of identifiers. When my partner and I go grocery shopping and split up to "divide and conquer," I do not have to perform a recital of his physical identifiers to be able to pick him out from the crowd. I have an unconscious and powerful sensory awareness of how the cluster of traits come together in the whole that is my partner. Thus, when experts at snake identification state they can identify snakes at the genus or species level by "just knowing," they are describing a keen sense of familiarity marked by attention.

Because moderators are highly familiar with and attentive toward snakes does not imply that most or even many of the community members obtain that same level of identification proficiency. Rather, people are choosing to participate in a community in which they are exposed to other people who have that deep familiarity and attention. They are experiencing that there are other people who do not just look at and see *snakes* but see *different species* and even *individuals*. This exposure is impactful, and its impact is heightened by another dimension of identification, the social dimension.

Social Identification

A second aspect of identification relates more directly to the rhetorical understanding of identification offered by Burke. In a 1951 essay, Burke offered that "old" rhetoric could be summed up in one word, *persuasion*, with a focus on deliberate design, and that "new" rhetoric could be summarized as *identification*, which includes unconscious factors (203). In a *Rhetoric of Motives,* Burke (1962) elaborated on the necessity of identification as a prerequisite to persuasion: One must identify with in order to be persuaded by, and the unifying work of identification is necessary because each life is an autonomous—that is, divided—being. Importantly, for Burke, to *identify with* is a provisional and incomplete affair; one can never completely share another's perspective or position. However, one can share moments of *consubstantiality*—a mixed ontological state of being both joined and separate, of being in a "community of shared interests while remaining autonomous individuals" (Jasinski 2001, 306). Burke's notion of consubstantiality resonates with Foss and Griffin's *autonomous interdependence* and what I am calling *individuals-in-communities*, and the state of consubstantiality helps explain how community members of snake identification boards come to "see snakes as they [the moderators] do."

Community members described how the rhetorical ecology of the snake identification boards—consisting of a clear articulation of community rules, the decorum the rules established, the predictability of the group's and a post's format, and the manner in which rules were enforced—created the space for members to see snakes differently. Each community member who spoke to me about how the snake identification groups helped them commented on the way in which the boards were moderated. People expressed how the clear rules offered a comfort of both knowing what to expect from posts and what was appropriate to post. One member said that this was particularly helpful to her as a neurodivergent person who often experiences anxiety in social situations. Another member said he was initially put off a bit by the rules because he does not like people being told what is/is not appropriate, but he said he later came to appreciate the guidelines because they facilitated his learning in that he could expect an identification and not get taken down a rabbit hole of some side conversation.

Rhetorically, these clear rules establish explanations for the form of the posts to the snake identification boards. Community members see an image of a snake and know that the "next step" is to identify the snake in the photo. Burke describes that when people can anticipate a form, they will participate in completing the form (Burke 2012, 262). Indeed, this is why Aristotle (1991) considered the enthymeme—a syllogistic argument that is incompletely stated—the heart of persuasion. What makes the enthymeme so impactful is that people are called to fill in—to participate—the argument themselves. Multiple community members described to me how they came to participate in the form of these snake identification groups. Even members who never posted in the group described how they "gamified" their Facebook feeds: rather than going directly to the landing page of the snake identification group(s) to which they belonged, they would scroll through their feed, see a picture of a snake, and immediately try to identify the snake; if they could, they would try to name the species, and if not that, to identify if it was a venomous or nonvenomous snake. They would then click on the post to see how the snake was identified and go on to check if there were other posts they missed while they were there. In this way, the *form* of the snake identification groups encourages participation, even if that participation is not shared publicly in the social media network.

Beyond the clear form, though, the moderators were praised for their consistently supportive tone. Members expressed confidence in the ethos of expertise of administrators, that these were people giving them accurate information based on informed knowledge, *and* they also appreciated the manner in which that expertise was conveyed. It was important to people that never posted themselves that *other community members* were never shamed. Members shared that even when rule violations needed to be addressed, the members appreciated the nonjudgmental tone in which the moderators addressed them. One member said that she came to the group for a snake identification and stayed in the group because of the people. In addition to retaining people who might otherwise come to the group once for an identification and then leave, the tone of the group led some members to internalize the pro-learning messages they encounter within the group. In some of the communities' group rules, it is explicitly stated that people's comments are to be treated with respect even if they say something wrong (the prime example being when someone says "poison" instead of "venom") because they are learning. In interviews, many members reflected that attitude to me. For example, one woman described to me how she liked to quiz herself "when posts are posted. I don't always get it right, but that's OK because I'm learning."

These experiences speak to the importance of *how,* the means by which someone is engaged in learning. The overwhelming positive regard that members had for the moderators enabled them to feel part of a common community with the moderators and to want to learn "to see snakes like they do" (Charland 1987, 133). An important part of the *how* is *identification*, in which identification functions as a means constituting the ends (i.e., that the creation of a space for understanding through these interpersonally competent means is also the end or goal state). George Cheney (1983) enumerates three typical means of identification: common ground (shared interests and values), antithesis (common enemy), and transcendence (moments of experiencing a collective we). What the snake identification groups offer is the means of identification of shared practice, a shared practice that exists prior to finding common ground (i.e., before members like or respect snakes, they learn to identify), in absence of antithesis (i.e., no one is allowed to be made a common enemy on the board, even people who post a picture of a snake they killed), and as a precursor to feeling part of a collective we. It is a simultaneously autonomous and collective experience of identifying, of a shared communal activity of looking at the image of a snake with intention.

This means of identification via shared practice is also seen in a paper by Tema Milstein (2011) appropriately titled "Nature Identification: The Power of Pointing and Naming." Pointing and naming, she writes, is a basic practice of using communication to discern the more-than-human world. Milstein argues that the shared practice of pointing and naming is an individualizing activity that allows people to see animals differently: as unique, as complex, as intrinsically valued subjects. "Pointing and naming can be seen as the basic entry to socially discerning and categorizing parts of nature. In this way, acts of pointing and naming generate certain kinds of ecocultural knowledge that constitute aspects of nature as considered, unique, sorted, or marked" (4). As discussed earlier, this is the work of making nonhuman animals visible, to get the animal to *show up* and *show up differently* in people's lives—to set "apart the named individual from the whole" (4).

Milstein's concerns about the shared practice of pointing and naming, of identifying with a particular nonhuman animal, is that the broader ecological context may get lost in the individual identification. This danger is particularly salient when the purpose of pointing and naming is to inspire pro-ecological attitudes and actions, as in the case of the whale-watching tours in which Milstein participated. Her concern is that identification can emphasize the individual at the expense of the community; therefore, she argues, one should work with the restorative potential of seeing each individual animal as intrinsically valued while weaving their stories into ecological contexts.

What makes the communal practice of snake identification different from pointing and naming whales is that snakes are not "charismatic megafauna" (Kleiman and Seidensticker 1985). The members I spoke with did not engage in practices like anthropomorphically naming snakes they encountered. Rather, they named the snakes according to species membership (e.g., "This is a rat snake"), and that species membership evoked ecological associations. For example, one member described how a neighbor was about to kill a rat snake, and she told the neighbor that rat snakes are beneficial members of their ecosystem, that the rat snake would help control rodent populations and be a deterrent for venomous snakes taking up residence in the area. Snakes' "roles in the ecosystem" was brought up by many members with whom I spoke. Even members who expressed no desire to touch, be near, or perhaps see another snake described a respect for the relation of snakes to their communities: as meso-level predators, snakes both control and feed the populations of many other species.

Of course, any act of pointing and naming is, essentially, an anthropomorphic act. Humans are critters who point and name. Language is a primary medium through which humans process the more-than-human world. A taxonomic identification may be reductive and distancing in some respects; the taxonomic identification provides a categorization that organize—and stereotype—one's thoughts about the particular creature being encountered. For example, one snake educator I interviewed contacted me when she had her first experience being struck at by a snake. She was struck at by a ball python with whom she was working, and she said to me that the signs were there that the snake was going to strike, but she did not attend to them. She did not attend to the signs because *the snake was a ball python, and ball pythons are docile*. In this instance, the identification separated her from the individual snake in front of her.

However, it bears repeating that, in a Lacanian sense of humans always already being socialized in a world of language, humans are critters who point and name. The desire to escape language is tantamount to wishing humans had a different way of perceptually organizing and interpreting the world. Further, the desire to escape language may itself be an anthropocentric longing: a desire to escape the thing that humans imagine separate them from the "rest of the animal kingdom" instead of recognizing that every animal has its own perceptual world, or *umwelt*. There is no state of unity that all other critters exist on from which humans are separated by language; language is just part of being a human critter.

What I want to put a fine point on here is that, through pointing and naming, individual snakes in ecosystems are being made salient for members of snake identification groups. That is, members describe both a sense of caring for the fate of "some snake six counties over" *and* an appreciation for the ecological niche that snakes fill. Furthermore, the communal aspect of shared practice *can* be accomplished through digital networks. Not every member needs to see the *same* snake "in-person" to share in the practice of looking and identifying. What this accomplishes is allowing people's attitudes and values toward snakes to transform *at a distance* (Peters, Sprenger, and Vagt 2020). Although the transformative potentials of these embodied encounters with snakes have been explored elsewhere (Rosenfeld 2020, 2021), it is important to recognize that snake identification groups are able to afford transformation prior to in-person encounter. Some members I spoke with expressed disappointment that they have not seen a snake since joining a snake identification group. They're "ready" now, and they want the opportunity to experience snakes differently. The ability of snake identification groups to enable transformation prior to an in-person encounter is an important component of the practice of snake identification.

Transformation of Individuals-in-Communities

Neurophysicist Donald Hebb wrote the oft-cited phrase, "Neurons that fire together wire together" to describe how neural pathways in the brain are formed and reinforced through repetition (Keysers and Gazzola 2014, 2). With repetition, neural networks—or neural circuitry, the way neurons communicate with one another—grow stronger. The brain creates metaphorical maps, and these maps help animals respond quickly to situations. The deep grooves of oft-repeated neural networks help explain why children with adverse childhood experiences experience physical, psychological, and emotion effects across their lifetime (Reading 2005). One way to understand this concept is to imagine going into a dense forest and taking a walk to a stream. The path traveled most often is clear, worn, easy to traverse, and affords a quick and easy (certainly easier) route to the stream. Although it would require a great deal of energy to choose and forge a new path, it is possible to take a different route, and neuroplasticity refers to the brain's ability to modify, change, and adapt both structure and function in response to experience throughout one's life.

However, when it comes to people's feelings about snakes, it can be hard to create new pathways. As described earlier in this chapter, there are a host of naturecultural factors that reinforce an aversion to snakes. Desensitization is a common form of psychological treatment for fears and phobias and involves extinguishing a particular emotional response (e.g., fear) from the stimuli that provoked the emotion (e.g., the image of a snake). In psychological settings, clients are typically taught techniques for relaxation while being exposed to less and less abstract versions of the stimulus (e.g., a drawing of a snake followed by a photo of a snake followed by an encounter with a snake in an enclosure followed by, perhaps, touching a snake). There are many reasons why a person may not choose to enter into a formal, therapeutic relationship for a snake phobia, such as the stigma of seeking assistance or the stigma of being afraid, the cost of therapy, and not having the time to devote to that kind of work, to name a few.

The participants I spoke with expressed their appreciation that they could access the snake boards on their own time, that their physical and emotional reactions could be private, and that they had control over how much or how little time they spent engaging with the group. For example, one participant—who works in the psychiatry department of a hospital and joined the board to engage in self-directed desensitization—told me that the first couple of days after joining the group were particularly challenging. "I would throw my phone across the room," she said. Despite the intensity of that initial reaction, she was surprised how quickly her decades-long phobia abated. "A couple of weeks," she said. She was not the only participant who expressed astonishment at "how little it took" to overcome something that once seemed insurmountable.

Photos are posted every day by people taking quick shots, often with their phone's camera, of snakes as they encounter them. Further, as there are snake identification groups for many states, community members in the United States often see photos of snakes in familiar landscapes. They see snakes in places where they *could* be—providing a sense of presence—but where they *are* not—providing a sense of absence; this sensation of presence-absence can be described as an experience of *liberated embodiment.*

Liberated embodiment refers to the idea that cultural texts, like film or in this case Facebook groups, allow people *embodied* engagement with an artifact because "it involves brain areas, such as the premotor cortex, that are also active when our bodies engage in actual movement" while also being *liberated* in the sense that "the simulation is offline, so to speak" (von Mossner 2017, 52). That the experience is digital does not make it any less real—that is, any less embodied and affective. Alexa Weik von Mossner (2017) describes how film viewers experience liberated embodiment: rarely does a viewer assume that "they *themselves* are physically present in the fictional storyworld of the film" but rather the use of subjective shots "facilitates *a sense of presence*" (60). The affect that people experience when they see the photos of the snake is real but so is their presence in the safety of their home; the simultaneity of these experiences are a productive affordance of digital rhetorics.

Another aspect that participants described as being beneficial to the desensitization process was the *unpredictable predictability* of encountering snakes on their feeds. *Unpredictable predictability* refers to the experience afforded by the interplay of the groups' clear community rules (the predictable) and the unknown of Facebook's algorithms (unpredictability). While much of the unknown of Facebook's—and other media conglomerates, like Google—algorithms are a cause for concern (e.g., see Noble 2018; Zuboff 2019), in this circumstance the unknown of the algorithm facilitates desensitization work. (Of course, this neither excuses nor justifies any company's obfuscation of how their algorithms work and their consequences.) As one participant described, she is motivated to check her Facebook because she wants to see what people in her community are sharing. By clicking "follow" on a snake identification group, her feed becomes interspersed with images of snakes. She does not know when these images will appear or how many there will be. Although this made the experience of going on Facebook anxiety-provoking for a while, she credits the surprise element with helping her change as quickly as she did. Like in the offline world, in which a snake sighting is usually unexpected, she was able to have a *liberated embodied* experience of those *unpredictably predictable* sightings.

The unpredictable-from-the-user's-perspective algorithm's[2] production of unexpected snake sightings was complemented by the predictability of the community standards of the board. As described, people very quickly learn the form that posts to snake identification boards take: An image will be posted, the snake will be identified, identifying features may or may not be shared, and the commenting will be closed. Likes are allowed, but members are told not to use the "sad" or "angry" emojis. The comfort of this predictability showed up in surprising ways in some members' lives. A few members expressed finding the snake identification board to be a place they sought refuge during the often-heated conversations taking place on social media during the 2020 US presidential election. Others shared that the snake identification community is the primary motivator for keeping a Facebook account.

Liberated embodied experiences of unpredictably predictable snake content afford new experiences with snakes *offline*. Some members describe that they still get startled when they see a snake in a body-to-body encounter. The startle is akin to an affective response, an intensity registered directly by the body and operating on an asignifying register. "Intensity is embodied in purely autonomic reactions most directly manifested in the skin—at the surface of the body, at its interface with things" (Massumi 2002, 25). The startle is the registration of the interface with a snake. After or simultaneous with the startle comes the emotional qualification, which is conscious and carries valence and meaning. For many, the startle remains but the subsequent qualification is profoundly changed. Before joining a community, members described being startled and then being filled with fear, helplessness, and sometimes embarrassment; now, the startle was often followed by something else entirely: "Oh! You're gorgeous," as one participant exclaimed to a rattlesnake who startled her on the porch.

A different community member, when asked what most surprised her about her experience in the snake identification community, replied, "I didn't expect them to be gorgeous. And they are." The aesthetics of snakes, like affect, are part of the asignifying rhetorical ecology at work in participants' transformation. The aesthetic is "always and everywhere rhetorical—that is, productive of effects—and crucially, these effects are produced on and through the live and lively bodies" (Hawhee 2009, 13). However, aesthetics are only part of this rhetorical ecology. More illuminating than the new aesthetic appreciation of snakes is to see what accompanied these evaluations: the change in individual lives.

"It's *liberating*," one participant said. "I can go out and explore." Other participants expressed a similar sense of new-found freedom, a freedom accompanied by a great sense of personal pride. Participants expressed astonishment at their journey. One participant described how she went from running in the opposite direction of snakes to having her husband call her to come outside to identify snakes he found in the yard. Another described stopping a group of people from killing a snake on a trail. Many shared the experience of having friends and family send them photos for identification or feeling empowered to intervene when incorrect identifications are being made on NextDoor. They expressed the pride they took in people asking how they knew so much about snakes, and the pride they took in putting in the work to learn. They conveyed a sense of confidence in their movements; "I know about strike distance. I know how far to stay away," one person said. Members expressed understanding that snakes were just "trying to live their lives" like everyone else and that being able to identify gave them "the confidence to live and let live." One person described how *exciting* it was to be able to accept snakes: "It is mind-blowing how much a person can change. It is mind-blowing the level of comfort and acceptance I feel."

The snake aesthetics, the sense of community, and the feelings of self-efficacy are all part of a rhetorical ecology. As the woman startled by a rattlesnake on her porch put it, "I am so proud of myself and so surprised at how gorgeous this snake was." It is a powerful rhetorical ecology that impacts moods and attitudes and, importantly, actions—and therefore lives. Joy is an expansive relation that increases one's power of living, and, it would seem, increases the power of living for others. This is perhaps best summarized by one interviewee: "I am so glad I could overcome my fear. In my own space, I am living really well with snakes."

A few of the people with whom I spoke joined the snake identification group to learn how to identify snakes around them, snakes of whom they were not afraid; however, the majority of people with whom I spoke described an intense hating or loathing borne from their fear prior to their time on snake identification boards. One described skipping "S" in the encyclopedia because she did not want to see the picture of a snake, another described being unable to watch film depictions of snakes for fear that the snakes could come through the screen (admitting, of course, that this was not rational), and others expressed not wanting to go on hikes or even being in their own backyard because there might be a snake. Some people had traumatic memories associated with their fear, like watching a snake be violently dispatched by an adult and then associating the snake with a memory of seeing a parent terrified and violent. Other people described just "always being afraid," having no memory of not being afraid of snakes. Whether people saw their fear as "nature" (an innate fear they always had) or "culture" (induced fear by the reactions of others or depictions of snakes), people still came to these boards with the belief that whatever the origin of the fear, things could be *otherwise*—that is, even if the fear was always present, that did not mean that the fear had to be permanent. The two common reasons that people gave for joining snake identification groups was either wanting to reduce their fear so that they could enjoy playing outside with their children or wanting (as one interviewee said it) to give themselves the gift of not being so afraid in their daily lives.

A common sketch of life prior to joining a snake identification group looked something like this: Whatever the origin of the fear, the presence of snakes was associated with an intense feeling of vulnerability to the point of helplessness. People felt unsafe in their yards, some even in their homes. Even as one community member was in a car about to run over a snake, she felt she was the one in danger. It may sound rather simplistic to claim that these were *bad relations* with snakes, but these bad relations—as Deleuze (1988), drawing on Spinozan ethics, describes—are relations that decrease one's power to act, one's power of living. Interviewees described how their fear of snakes restricted their lives and their ability to participate in communities of importance to them; their individual lives and their communities suffered. It is evident from these accounts how these human-snake relations were poisonous (in the sense that Spinoza uses the word, as discussed in chapter 2) for humans and snakes, decreasing the *power of living* for both. Humans were living in utter terror for their own and loved ones' wellbeing, and snakes were dying from this terror. "What poisons life is hatred" (Deleuze 1988, 26).

However, the converse of poison are tonics or remedies, and remedies restore the body. It is worth noting that a substance is not innately a poison or a remedy; the relation matters. Even a snake's venom, which can and does cause the death of humans, can also be a remedy for thrombosis, arthritis, and other illnesses (Pal et al. 2002). Remedies are *good*, in that they increase the power of living; they increase joy. And joy is an expansive relation—one that nurtures individuals and affords communality.

What the snake identification groups do is offer a remedy. People come to these groups from a place of vulnerability—vulnerable because they are living with fear, vulnerable because there is concern of saying or asking something "stupid"—and through the group's invitational rhetoric, they are offered tools to help them flourish. The instantiations of social identification that take place in snake identification groups contribute to what Chris Ingraham (2020) calls an *affective commonwealth* (159). Ingraham draws attention to the community-building capacity of gestures (i.e., the means by which people engage in public life), even small gestures like the sending of a get-well card. The sender of the get-well card understands that the card does not carry medicinal properties. Rather, such gestures are affectively generative and have the potential to create a sense of connection and interdependence.

Members of the snake board do not only share their stories with me. They share their stories with each other. Indeed, the reason I even knew to seek interviewees about the transformative experience of snake identification groups is because as a longtime participant-observer of the board, I had read several posts attesting to how lives had been changed. When I asked the administrators of one board if I could solicit participants on the group's feed, they said yes *and* sent me stories they had collected over the years. Interviewees also shared the joy of reading others' stories of transformation. The sharing of stories is a gesture of expansive joy—sharing with others the possibilities of living differently. Just as there is joy watching others play (Bekoff 2007, 95), there is joy in watching others live fuller lives.

A READINESS TO LIVE WELL WITH NONHUMAN ANIMALS

In this chapter, I explored how digital networks can function rhetorically to transform ways of living in and with the more-than-human world. Online snake communities and offline snake encounters can have a synergistic rhetoricity that challenges anthropocentrism and transforms aesthetic evaluation and ethical engagements with snakes. Community members of snake identification groups described how the people and processes of the groups afforded the change from feeling *vulnerable to* snakes to feeling *vulnerable with* snakes; that is, they were able to recognize and act on the belief that both they and the snakes were just trying to live their lives. Where *vulnerable to* was often described as a helpless state marked by fear and hatred, *vulnerable with* was a state of flourishment—a state of "living really well with snakes"—marked by the recognition of shared embodiment and mortality.

Yet, while these communities enable this kind of cross-species flourishment, the embeddedness of these communities in larger social communities, like Facebook, reveals some of the vulnerabilities, as well (i.e., the disenchantment with a platform may overcome the desire to remain in the group). Even when community-care is flourishing, as it seems to be in multiple social media snake identification groups, that flourishment is interdependent on other communities that may not be performing caring work adequately. That multiple interviewees shared that the snake identification groups were the only thing keeping them on Facebook speaks to the fragility of positive rhetorics that are embedded in social media platforms, as well as to the fragility of ecological readiness.[3] Ecological readiness is not an on/off switch but a spectrum of capacities, intentions, and practices that are impacted by rhetorics and infrastructures. In accordance with the community-care paradigm, an *ecologically ready* participant is not an atomized individual but a member of communities that enable and constrain certain kinds of participation and even subject positions.

I described in this chapter how invitational rhetoric helps explain the rhetorical efficacy of the snake identification groups through the ongoing attention to individuals-in-communities. As I conclude, I want to linger a moment on the significance of an invitational rhetoric that attends to individuals-in-communities. Scholars across the humanities and science disciplines have studied the centrality of *belonging* to flourishment. Psychologist Matthew Lieberman argued that Abraham Maslow's famous hierarchy of needs for survival—which places safety at the bottom of the pyramid, belonging in the middle, and self-actualization at the top—is wrong about the most fundamental need; social bonds are the most essential needs of humans for survival (Lieberman 2013).

The community members of snake identification groups provide insight on how an expansive network of belonging can flourish. People coming to the board are aware that the administrators and moderators of the board like snakes and want others to coexist with snakes. They are aware that there is hope that people will *live differently* with snakes. However, the sense of community that people perceive suggests "that intent is only one element in a large array of things, feelings, peoples, and forces all complexly interacting" (Rickert 2013, 36).

Snake identification groups show how new, worldbuilding rhetorics for encountering snakes in one's life are possible. Snakes can *show up differently*, and once they show up differently, relations can transform. For many members of snake identification boards, a snake's presence may still startle. But after the experience of being startled, the emotional response is malleable and subject to change. After the moment of separation—the division of one's self from the environment marked by the startled response—there can be identification of and identification with the more-than-human world. That is why the possibility and processes of transformation still matter in thinking about the Anthropocene because in many ways, to paraphrase Spinoza, we do not even know what an *anthropo* can do or what (more-than-)human relations are possible. The next chapter, on *infrastructuration*, will show more fully that there are many ways of living as humans in the (more-than-)human world.

STORY TIME: BECOMING A CHICKEN PERSON

"It's ridiculous. I can work with any dog, but the chickens terrify me," one of the seasoned volunteers said to me.

I was deployed as part of a rescue team taking care of the animals rescued from one of the largest dog-fighting rings in the United States. There were 367 dogs confiscated, and a few chickens. I was not officially part of an animal rescue team, but I was an intern with an animal advocacy group that was assisting in the rescue efforts. The case had gone on so long that the usual rescue teams were exhausted, and groups were scrambling to get more people to help provide the care needs.

It was a very different situation for me: tickets were booked for me, and I did not know where to report until the last moment. I was not to share locations with anyone. The animals that I was going to be helping care for were evidence in a large multistate legal case, and it was possible that there could be an attack on the makeshift shelter if the location got out. It was summer, and it was very, very hot. The day started before sunrise, and the work lasted until the evening.

For the week that I spent assisting in the care of these animals, I was stationed in the building that had the chickens. The chickens were not a favorite among many of the caregivers. When someone came in their enclosure, they would rush toward the person, jump, and fly up to try to get the food being carried. It was startling. It was also amusing; it was funny that the chickens would be the intimidating creatures in this situation. I got into it: going into their enclosure, dropping the food, and picking up the eggs as quickly as possible, and then hurrying out and shutting the gate behind me before letting out a "whew," like one might see in a monster movie.

A couple of days into my visit, one of the veterinary technicians pulled me aside and asked if she could show me how she interacts with the chickens. She did not ask in a manner that ridiculed, nor was she demanding that I watch and learn. She genuinely asked if I would be willing to participate with her. She said that she knew that there was some nervousness around the chickens, and she understood why. However, the nervous responses were just fueling the situation, and she wanted things to go differently—for the chickens and for the volunteers.

I went into the enclosure with her, and I watched her gently lower herself to the ground. In hardly a moment's time, chickens were coming over to her and examining her. I watched her pick one up and hold the hen in her lap. The chicken pecked a bit at her watch, and she told me that the chicken was just curious and that those curious, exploratory pecks don't really hurt. She spoke softly to the chickens and to me. This moment also felt like something out of a movie, but this time it wasn't the horror genre. This moment felt magical, like I was watching some fantastical moment with a Dr. Doolittle–esque character who could talk to nonhuman animals. But it was even better than that because it was not magic. It was the transformative potential of new relations—relations that can begin at any moment.

Instead of fear and nervousness begetting fear and nervousness, a way-of-being was created in which I could move differently around the chickens, and they could move differently around me. Relationships are always capable of transformation. I am never more aware of this capacity than seeing documentary footage of "predators" and "prey" gathered around a shared watering hole. Relations are not static; one does not exist in a concrete position to another being. There were many ways the chickens and I could relate to one another, and frantic encounters were only one of those possible relations.

She asked if I would be willing to spend a few minutes every day just being with the chickens, without bringing them food, so that they could start to experience people differently. Not as frantic creatures dropping food and fleeing, nor whatever they had experienced prior to their confiscation, but as another *being* just living alongside them. The day I accompanied her into the chicken's enclosure, I called my partner to describe how wondrous this technician was. The next day, I called my partner and said that I could not imagine finishing my life without sharing it with chickens.

I tell people I became a chicken person on that pivotal day, but the reality is that was only the launch of the ongoing work of *becoming a chicken person.* As the other stories in this book make clear, every bird and flock configuration invites me into human-chicken and human-flock relations and activities. Many days, *becoming a chicken person* means living calmly alongside a group of birds to whom you provide food, water, and a clean shelter. Some days, *becoming a chicken person* means working frantically to determine and try to treat a bird's illness. I did not know what sour crop was until I nearly lost a pullet to it. I do not know what *becoming a chicken person* will mean tomorrow, or even later today, but I look forward to finding out.

NOTES

1. Levin (2016) cites Sweetwater, Texas, Mayor Greg Wortham as saying the rattlesnake roundups are “part of our identity, our community fabric.” To speak against rattlesnake roundups, then, is a political act and an enactment of what Eva Haida Giraud (2019) calls an *ethics of exclusion*, in which taking a stand for particular relations necessarily means excluding others. The thrust of an ethics of exclusion is the acknowledgement that all engagements are noninnocent and entail exclusions, *and* people can take a stand for certain relations and certain worlds in an ethical manner by recognizing the contingency of the position being taken, making one’s commitments and the associated exclusions visible, and leaving open the possibility for things to be otherwise.

2. Although the details of how Facebook’s algorithm selects what to (not) show are unknown, it is known that the algorithm evaluates and scores posts in interest of each user to maximize each user’s time on the platform. The algorithm has evolved over time, initially working more simply to bump posts with the most likes to more recently including things like time spent, weighting reactions, completion rate of videos, relationships, etc. (Newberry 2022). The algorithm is a significant agent in the effectiveness of the snake identification boards. It is possible that, as the algorithm continues to evolve, the snake boards will have less of a presence in people’s feeds and thus have diminished rhetorical force.

3. While many of Facebook’s features enable group moderators to enact the kind of invitational rhetoric that community members appreciate, people also expressed disenchantment with the platform as a whole.

The analysis of snake identification groups provokes the question of what an invitational social media *platform* would look like, and the analysis also helps envision that possibility. An invitational platform would—at minimum—be transparent, comprehensible, and responsible to community members. The terms of service (ToS) of Facebook, the platform hosting the snake identification communities, are long and difficult to parse. For example, the third point under Section 5 “Other” of Facebook’s ToS reads:

> If any portion of these Terms is found to be unenforceable, the unenforceable portion will be deemed amended to the minimum extent necessary to make it enforceable, and if it can’t be made enforceable, then it will be severed and the remaining portion will remain in full force and effect. If we fail to enforce any of these Terms, it will not be considered a waiver. Any amendment to or waiver of these Terms must be made in writing and signed by us. (Meta 2022)

It is unclear what this statement means for an individual user. Facebook’s ToS would be unlikely to pass as a consent form submitted to an Institutional Review Board, and it certainly does not establish the clear and coherent community standards that members valued in snake identification groups. That groups on Facebook are able to craft invitational spaces speaks more to the inventional capacities of individuals-in-communities than to the norms established by the platform.

Chapter 6

Infrastructuration

So far, I have discussed the significance of disclosure and transformation for building more-than-human worlds of "earthly coexistence" (Barnett 2019, 287). Disclosure (how things show up) and transformation (the capacity of things to change) are involved in worldbuilding rhetorics that seek to live well with(in) the more-than-human world. Now, I return to the idea of worldbuilding notion in chapter 2 from *Gorgias*: that rhetoric's fullest impact can be seen in the bridges connecting places. The bridges' construction were undoubtedly a consequence of rhetoric, but bridges are also rhetorical agents of their own. Bridges can connect temporalities (e.g., as historical landmarks), please or upset aesthetic sensibilities, and serve as a means of mobility that can afford or limit health (Kirsch 2013). Bridges also involve ethics of care: how they are (not) maintained, what lands they connect, how they connect those lands, and to what consequences for which communities. Bridges also provoke questions about what a land ethic entails for highly mobile societies and how different infrastructures afford different mobilities (e.g., walking, biking, driving) that disclose and transform the landscape in their own, particular ways (Kirsch 2013).

How do different infrastructures disclose and transform the landscape in their own, particular ways? This chapter explores how the things (more-than-)humans build are made possible by existing thoughts, materials, practices, beliefs, and attitudes (i.e., digital, material, and embodied infrastructures) *and* simultaneously how they become infrastructure for emergent thoughts, materials, practices, beliefs, and attitudes. To stay on this road, I trace what I am calling *infrastructuration*—the simultaneous digital, material, and embodied worldbuilding processes of making and being made; the simultaneity of creating/shaping/arranging and being created by/shaped by/arranged by—by first engaging with science and technology studies and informatics (STS) to explicate how infrastructure is rhetorical, and then surveying some of the infrastructure of rhetoric through a reading of *memoria* and *topoi*. Next, I look at the rhetorical and ethical consequences of two worldbuilding projects' infrastructural elements—the Pink Chicken Project and Reclaim the Void—and then offer some concluding remarks on how these projects expand thinking on *worlding well*.

BECOMING INFRASTRUCTURE

It will help to begin with a working understanding (as opposed to a definitive definition) of *infrastructure*. Bowker and colleagues (2010) describe infrastructure as "pervasive enabling resources in network form" (98). This conceptualization is purposefully broad, allowing nontraditional practices and materialities to be understood as infrastructure. The significance of this conceptual widening is that it more fully accounts for individuals, materials, and daily practices that build, enable, and maintain processes, communities, and systems. For example, think back to the account of the bridges of Athens discussed in chapter 3. Bridges are clearly understood as infrastructure. However, one can also understand the relationships and discussions that lead to the building of bridges as part of the pervasive, enabling resources that allowed the bridges to come into being.

Studies of infrastructure foreground the processes of distribution and the unique materialities of distribution, which include labor, technologies, resources, and relations (Parks and Starosielski 2015). The last term in the list is particularly important. Some infrastructures may be made of concrete, like a sidewalk, but infrastructures are also profoundly relational (Star and Ruhleder 1996, 112). Infrastructures have degrees of stability and may themselves be stabilizing, but they are certainly not concretized. Infrastructure "encompasses both technical bases and social arrangements, extends beyond single events and sites, connects with existing practices and standards, and must be learned and naturalized over time by users" (Parks and Starosielski 2015, 9). The learning and naturalizing over time—the *becoming infrastructure*—is deeply entangled in rhetorical, ethical, and political concerns, and I elaborate here on that idea of *becoming infrastructure* through the concept of *infrastructuration*.

Infrastructuration is a neologism to refer to the dynamic event of worldbuilding with embodied, material, digital, and symbolic elements. STS already has a productive term for studying the processual, in-the-making aspect of infrastructure called *infrastructuring* (Bowker et al. 2010; Karasti, Pipek, and Bowker 2018) and some affiliated methods for making emergent infrastructures visible. Such methods include practical methods of observing infrastructure during breakdown (Star 1999), such as observing a highway being repaired, and conceptual (see Bowker 1994) and ethnographic (see Simonsen, Karasti, and Hertzum 2020) methods of "infrastructural inversion" that shift attention from the activities supported by infrastructure to the activities that enable infrastructure to function). Infrastructuration does not replace but supplements *infrastructuring*.

Coming from the Latin ātiōn, the suffix *ation* indicates an action (e.g., the act of altering is alteration), condition (e.g., the condition of being starved is starvation), or result (e.g., the result of compiling is a compilation). What is productive about infrastructuration as a concept, then, is its ability to afford a multioptic view of the actions, conditions, and results—or consequences—of the processual, in-the-making aspect of infrastructure. As a concept, infrastructuration is akin to the subject-in-the-making term offered by Stephen Wiley and Jessica Elam (2018)—*synthetic subjectivation*—in that both terms note the dynamic, recursive processes that both *result from* and *intervene in* a composition of embodied, material, digital, and symbolic elements.

Bear with me for a moment while I describe the process of infrastructuration through the idea of the *adjacent possible*. In *Where Good Ideas Come From*, Steven Johnson (2011) describes how evolution and innovation occur primarily in the realm of the *adjacent possible*—in other words, the realm of possibilities available at a given time and space. For example, evolutionarily, human life was not possible on Earth four billion years ago. Innovatively, Amazon—as we know it today—could not have happened during the days of the telegraph. It is, of course, conceivable that someone along the lines of H. G. Wells could have imagined such a corporation in the nineteenth century: "[E]very now and then an idea does occur to someone that teleports us forward . . . skipping some exploratory steps in the adjacent possible. But those ideas almost always end up being short-term failures, precisely because they have skipped ahead. We have a phrase for those ideas: we call them 'ahead of their time'" (Johnson 2011, 36).

Johnson explains the adjacent possible through use of a spatial metaphor: a house that (magically) expands with each door a person opens. A person enters a room with four doors to four unvisited rooms. Each of those doors is an adjacent possible. When one opens a door, more doors—more possibilities—appear. An idea "out of time" would be a door without the material infrastructure to access it; it is not an adjacent possibility because it is unreachable. But, Johnson says, "[k]eep opening new doors and eventually you'll have built a palace" (31). This ever-unfolding, highly relational palace is important, and I will return to that shortly, but before I do, I need to bring rhetoric more centrally into the discussion of infrastructuration and elaborate a bit more on the rhetoricity of infrastructure.

From a rhetorical perspective, infrastructuration refers to the process(es) that invent those pervasive, enabling resources in network form. As a concept, it is informed by Foucauldian sensibilities of questioning the conditions of existence of phenomena and media materialist theories that examine how media function. To study infrastructation, then, is to study the (un)making of places, practices, and attitudes (Towns 2022). Infrastructuration is not in itself a testament to the durability of the infrastructure (i.e., the concept is not analytic), nor is it an endorsement of the ways-of-relating that the infrastructure affords (i.e., the concept is not ethical). Rather, the concept opens onto questions of durability and ethics through its attention to the "irreducible rhetoricity" (Davis 2014, 537) of enactments and objects to occasion consequence and forge/forego certain connections.

As described in chapter 1, there are many concepts in rhetoric that illuminate how rhetoric is part and parcel of worldbuilding (e.g., ambient rhetoric, procedural rhetoric), which include the invention and maintenance of infrastructure. STS and informatic studies also have noted the rhetorical nature of infrastructure, describing how rhetorical mechanisms afford and constrain how people view infrastructure (e.g., what metaphors are used to describe and understand infrastructure), how the existence of infrastructure is always the outcome of debate, and how design makes some of infrastructure's consequences more visible than others (Bowker et al. 2010, 98–99).

Infrastructuration is related to circulation, in that both infrastructuration and circulation are concerned with the movement of discourse, bodies, artifacts, and ideas and the constitutive force of these movements in cultures and communities. However, where circulation refers to "a cultural-rhetorical process" of material and discursive movement and transformation that impacts various ecologies, communities, and subjectivities in networked environments (Gries 2018, 12), infrastructuration refers to the process(es) of creating media, channels, places, abilities, and literacies that enable or inhibit circulation. Although the digital sphere makes the communicative and transformative process of (re)circulation particularly visible, the "underlying structure[s] of possibility" (Jackson 2018, 235)—from the algorithms to the server farms to the maintenance workers in the offices tending the fiber optic cables to the more-than-human earth from which the minerals to create screens and coils are extracted—is largely invisible. Neither this chapter nor any book will (or can) fully reveal the underlying infrastructures of the digital realm. What this chapter does, though, is open to a more expansive understanding of digital infrastructures beyond the opaque—or blackboxed (Latour 2000)—codes and technologies to include the histories, politics, and ethics that underlie what circulates in digital spaces. An expansive articulation of digital infrastructures is itself a method of resistance to technological determinism; it foregrounds the materiality of digital technologies and embeds them in (more-than-)human affairs and undermines any "just so" stories of how technologies act in the world.

What is and is not infrastructure is not known *a priori*. Memes, well-suited for studying circulation, may also be infrastructure. In other words, what circulates "today" may become infrastructural "tomorrow." The thrust of the concept of infrastructuration—as I am using it—is not to get into some ontological weeds about whether a particular artifact, for example, is circulating or becoming (or both) infrastructure at a given time; rather, it is to provoke and animate questions about the consequences of that artifact becoming infrastructure. In other words, *infrastructuration* is a speculative inquiry into the kinds of worlds an artifact participates in building. For example, the Latourian speedbump participates in building a world in which automobile traffic is slowed *and* permitted.

With this articulation of the rhetorical nature of infrastructure, I now want to return to the memory palace and discuss some aspects of the infrastructure of rhetoric, which are also—in turn—worldbuilding concepts themselves: *memoria* and *topoi*. Johnson's ever-unfolding palace is helpful here. In the study of rhetoric, palaces bring to mind the ancient tradition of memory palaces—that is, techniques for the accurate and efficient retrieval of information. Memory palaces provided the mental infrastructure that allowed individuals to readily quote from epic poetry or even recall sociospatial arrangements after brief exposure. Memory, or *memoria* (one of Cicero's five canons of rhetoric), encompasses episodic memory (specific memories with emotional, geographic, and temporal tags), semantic memories (generalized, abstract concepts), and predictive simulation (use of memory to project into the future). "Memory is an interrelated system of cognitive abilities that allow us to recognize, recall, associate, integrate, and define the data we gather through sensory perception, reason, and social learning" (Parrish 2014, 129). *Memoria*, both the kinds inscribed in minds and the kinds inscribed on/in bodies, is a kind of personal, rhetorical infrastructure that helps afford and simultaneously limit one's adjacent possibilities.

However, memories are not only individual. Memories are, of course, filled and fueled by relations, and there are also communal memories. Memories, too, exist as individuals-in-communities. For example, September 11, 2001, is communal memory for much of the globe, but there are different communities of memory—how the event is remembered and with different affective orientations. As a citizen of the United States, I share in a communal memory of 9/11 as a tragic and transformative date in the country's history, and I also have my individual memories of that day, which include hearing that the planes hit the World Trade Center while I was in physical education class in high school. That memory, of course, circulates in a network of memories that makes different aspects salient and inflected in different, and sometimes conflicting, ways. Further, my memories are not some concrete inscription of an event but rather pervasive, enabling resources that continue to affect and signify "across breaches in time and circumstance" (Vivian 2018, 294). Memories are not passive but are intersubjective, interactive, and dependent on "social frameworks of memory, or socially organized patterns of recollection in families, religions, and other groups" (291). An episode of Netflix's *Explained* had people share their most vivid memories of September 11 and found that many of those memories depict scenes that could not have happened as remembered. For example, one person recalls seeing smoke from the buildings from her classroom, but her classroom was not positioned in a way to enable such a view (Cole 2019). The memory of the event was inflected by the *enargeia*—or vividness—of the news, creating a sensorial memory that recalled more the intersubjective experience of the day than the physical facts of the day.

Thus, it seems we have individuals-in-communities all the way down. Again, though, it is productive to separate—even if artificially—the individual and community for a moment. If *memoria* serves as an individual (but nonetheless always already communal) infrastructure of rhetoric, then what serves as a communal (but nonetheless always experienced by individual subjects with local meanings and values) infrastructure of rhetoric?

On my reading, *topoi* are a communal infrastructure of rhetoric. Before I explain this, let me first acknowledge a strong and valid argument against *topoi* as communal, rhetorical infrastructure. This may sound backward: if I have not yet provided the resources to enable understanding *topoi* as communal infrastructure, how is one supposed to make sense of the argument against *topoi* as communal infrastructure? In this case, starting with the refutation will actually better serve the purpose of illuminating *topoi*.

Topoi are often translated from the Greek as "commonplaces" and theorized as resources for crafting arguments. Thomas Rickert (2013) displaces *topoi* in favor of *chora*. For Rickert, *topoi* as common*places*—that can be returned to—seems too static. In *chora*, Rickert finds a kinship term for animating his conceptualization of ambient rhetoric. Tracing *chora* through the works of Plato, Kristeva, Derrida, and Ulmer, Rickert finds in *chora* a term for place that gets at the emplacement of rhetorical invention. *Chora* are dynamic spaces of movement. I agree with Rickert that an inert understanding of commonplaces is not only insufficient but theoretically unsound. However, I find more in *topoi* than inert commonplaces. I think the commonplaces work more like the magical palace in Johnson's metaphor: offering communal resources for invention and also always being modified by invention.

Topoi, as theorized by Matt Pritchford (2022), can serve as places for arguments and also as means of disclosure. Pritchford traces the formal characteristics of topoi: (1) as commonplaces where one can look for resources for argumentation and invention, through Aristotle and Perelman and Olbrechts-Tyteca; (2) as a concept related to content through Finnegan, Hariman and Lucaities, and McGee; and (3) as particular functions through Michael Leff, Cintron, Olson, and Boyle. Bringing these perspectives together, Pitchford arrives at an understanding of *topoi* as practices of showing forth that are made stable through repetition. *Topoi* are practices of stabilization. Importantly, the stability can also be unmade.

Following Pitchford's perspective of *topoi*, *topoi* can be understood as communal, enabling resources of and for rhetorical invention. The stability and dynamism that Pritchford notes in *topoi* are also characteristic of Bowker and colleagues' (2010) conceptualization of infrastructure. To reiterate an important point, neither *memoria* nor *topoi* can be easily separated into individual infrastructure (as memories are always social) or communal infrastructure (as commonplaces are still accessed by individuals with unique perspectives). I am certainly not the first to note the linkage of *memory* and *topoi*. For example, Michael Leff (1996) reads the ancient scholar Quintilian as binding memory with invention through *topoi* (or *loci communes*). *Memoria*-in-*topoi*, infrastructure-in-infrastructure, individuals-in-communities: The making and being made of infrastructuration is always already occurring in complex networks.

The complexity of infrastructuration—a dynamic process occurring in entangled universe(s)—does not relieve us of the sociopolitical responsibility to assess and intervene in infrastructure. As Eva Haifa Giraud writes, "Although nothing might exist outside of relation, *certain* things might need distance from *certain* relations in order to allow particular realities to be enacted and preserved" (2019, 173, emphasis in the original). Engaging in sociopolitical issues requires a noninnocent carving of space for "possibilities for action amid and despite this complexity" (2). What can be done amid complexity is work to make practice more responsible and response-able: more visible and open to contestation.

The concept of infrastructuration provokes many possible lines of inquiry into the Anthropocene. There is the most obvious, "What are the infrastructures that afford the Anthropocene?" but this question has already been widely asked and answered: mass industrialization, including the industrialization of agricultural practices (Lin et al. 2011), globalization of inequitable capitalist-consumerist economic relations (Barber 2021; Josephson 2020), and energy-intensive mobilities (Mann and Kump 2015). I refer to these as *anthropocentric infrastructures*.

There are many questions that are productive in illuminating these anthropocentric infrastructures. How and to what consequences do these infrastructures invent subjectivities through their impact on *memoria*? How do these infrastructures invent communities through the *topoi* they afford and the *topoi* (e.g., metaphors) used to understand infrastructures? Where are the points of vulnerability in these anthropocentric infrastructures? Under what conditions do these anthropocentric infrastructures flourish? What cultural narratives and practices are remembered and enacted in/by this infrastructure, and what cultural narratives and practices are forgotten and marginalized in/by this infrastructure? How does this infrastructure invite individuals to navigate? How does this infrastructure invite communities to navigate? What infrastructure is needed to afford different ways of navigating? What kinds of dwelling with more-than-human others could other infrastructures afford?

I will not get to all of these questions, but in the remainder of this chapter, I will get to a few by first offering an account of (de)colonial infrastructures and then providing rhetorical analyses of two digital, worldbuilding projects. I then conclude by discussing how the insights gleaned from these two projects interact with the concepts of *disclosure* (chapter 4) and *transformation* (chapter 5) and help to envision what possibilities exist for reclamation of the *anthropos* in an age characterized by irreparable loss.

(DE)COLONIAL INFRASTRUCTURES

"Anthropocentric infrastructures" is both a shorthand for the pervasive, enabling, networked resources (ranging from neoliberal logics to carbon-emitting bodies) that afford the Anthropocene, and is also a misnomer. As Donna Haraway and Anna Tsing caution, the Anthropocene might more aptly be called the Capitalocene (Haraway 2018) or Plantationocene (Tsing 2015). The care one needs to take when using the term Anthropocene is to be mindful of and explicit in stating that not all humans and their associated ways-of-life contribute to current environmental and climatic crises. There are many ways of being *anthropos*, which happens to be the best news one could deliver. Humanity is not bound by some biological necessity to relate to the more-than-human world in ways that alter the conditions of livability in the biosphere. As Zoe Todd (2015) writes, "With the prevalence of the Anthropocene as a conceptual 'building' within which stories are being told, it is important to query which humans or human systems are driving the environmental change the Anthropocene is meant to describe" (244).

Todd explains how the Anthropocene can function as a *topos* through which stories get told and memories get made. I agree that the word Anthropocene does have this gathering effect. Further, Todd's example helps show how *topoi* are infrastructural. The Anthropocene is a *topos* that gathers complex outcomes of anthropocentric infrastructural arrangements, which include things like carbon-polluting industrial processes and the GDP as the primary measurement of how "well" a nation is. The stories, discourses, and effects that gather under the term *Anthropocene* are dependent on anthropocentric infrastructures.

Todd makes another point to which it is worth attending. Todd writes that she reluctantly uses the term "Anthropocene"—reluctant because of its catch-all nature—but uses it because it is a way to intervene in ongoing conversations about human-driven environmental change and to raise the important question of "which humans?" (244). In this respect, Anthropocene in scholarly literature works much like #Anthropocene—a tag to link perspectives, even if that perspective is dissenting within the conversation (de la Bellacasa 2017, 78), such as using the *Anthropocene* to critique the term's broad and unnuanced strokes. I now turn my attention to painting the Anthropocene with more finely bristled brushes by discussing (de)colonial infrastructures and their relationship to the anthropocentric infrastructures affording the Anthropocene.

Colonial Infrastructures

Colonial infrastructures have been well described by intersectional and anti-/decolonial theorists. At its broadest, colonial infrastructuration can be thought of as the discourses, practices, and material arrangements that separate and rank humans from each other and from the more-than-human world (Deckha 2018). Informed by decolonial theory, I avoid invoking a common trope: *the Cartesian divide*. The Cartesian divide erases the complex network of colonial practices and material arrangements that diminish and separate the more-than-human world from the human world. The philosophy of René Descartes—with its mind/body and attendant human/nature divides—provided important material for the co-construction and maintenance of colonial infrastructure, but it is only a part of the infrastructure. As discussed earlier with the concept of the adjacent possible, Descartes lived in material and ideological conditions in which his ideas could take hold. Even within colonial infrastructures, it is not that there were no counter-philosophies: for example, Voltaire famously chastised Descartes's human/animal separation as shameful. However, the colonial infrastructure better afforded the circulation of an ideology of human separation and superiority.

The discourses, practices, and material arrangements that separate and enable hierarchies are articulated with particular epistemological and axiological perspectives. Colonial infrastructures are partially built by the belief in and desire for bracketed and disembodied knowledge, what Donna Haraway (1988) calls "the god trick" (582), or epistemological rhetorics that claim to produce an objective view *from nowhere*. Colonial infrastructures tend to be built on ethics based on likeness, such as the *identity theories* (Calarco 2014) that bring nonhuman animals into human ethical frameworks to the extent that they demonstrate some important characteristic for moral consideration (e.g., nociception [the nervous systems' processing of noxious stimuli to produce a subjective experience of pain], rationality, sentience). In short, colonial infrastructures are built on and encourage *knowing* from afar and *considering* based on likeness.

Even as I attempt to nuance the portrait of anthropocentric infrastructures, there are two important points to make so as to avoid painting too broadly with the colonial infrastructure paintbrush. The first is to make clear that even within colonial infrastructures, like identity theories, there have been good and important political outcomes. For example, Peter Singer's (2009) *Animal Liberation* is often critiqued for its utilitarian ethics, the marginalization of differently abled people, and for its search for criteria of likeness (e.g., the capacity of humans and nonhumans to feel pain) to bring animals into human moral imagination. However, as Calarco (2014) takes care to explicate, works like Singer's have also done a great deal in bringing nonhumans into humans' moral imaginations and in mobilizing political action (e.g., campaigning against vivisection). Indeed, it is doubtful I would be writing this book without having read *Animal Liberation* in 2014 and finding my understanding of what counts as social justice substantially widened.

The other point is to make explicit that being born into a colonial infrastructure does not mean that one is bound to that infrastructure; it does mean that one is living and moving through systems in which hierarchies and separation pervade. To paraphrase Karl Marx (1972), people may make their own ethical commitments to relate to other humans and more-than-humans in more affirming ways, but they do not make those commitments in circumstances of their choosing (245).

To illustrate this point, consider Aldo Leopold: The author and ecologist was born in 1887, just a few years before pragmatist Charles Sanders Peirce (1992) would declare the nineteenth century the *Economical Century*, a time in which economics and evolutionary theory were merging to create a totalizing view of the world. Yet, Leopold still went on to develop his *land ethic* in *A Sand County Almanac* in the mid-twentieth century. Leopold's land ethic, as will be made clear in a moment when I turn to decolonial infrastructures, is a marked departure from the instrumentalizing view of people and the more-than-human world that Peirce documented decades before. This land ethic, as discussed in chapter 1, expanded the concept of "community" to include the more-than-human world—and not only nonhuman animals and plant life but mountains, soils, and waters. Leopold recognized that the flourishment (or lack thereof) of people and the land are intertwined. Akin to an ethics of care, the land ethic offered neither deontological rights of or duties to the land nor a utilitarian calculus to solve dilemmas when human and more-than-human needs are in conflict. Rather, Leopold imagined that care emerged from thick, rich involvement of lived experiences with the more-than-human world.

With those two points made, part of the work of decolonizing research is reaching beyond the Western canon in efforts to make room for land ethics—including multispecies ethics—and politics. Walsh (2010) and Todd (2015) express concern about the level of abstraction in which much posthumanist scholarship operates; it may be working to decenter the human, but it does not quite seem to situate the human in an active material world. Further, Walsh and Todd note that many of the realizations of posthumanist scholarship were always already realized by other humans. Thus, while I will not throw out the rich and caring insights of Leopold's land ethic, I would be engaging in colonial infrastructuration if I were to write as though the recognition of more-than-human entanglement began with Leopold (see, for example, Rose's [1988] account of Aboriginal land ethics). Leopold's perspective is a fellow land ethic, not a founder land ethic. More specifically, Leopold's perspective is a fellow ethic to Indigenous Place Thought and ethics of interdependency, to which I now turn.

Decolonial Infrastructures

Decolonial infrastructures refer to political and economic structures—such as the *buen vivir* post-development movement in South America—as well as emancipatory structures of thought and relations that are committed to ongoing undoing of the ontology and power apparatuses established by colonialism. Decolonial infrastructures make visible the always already embeddedness of human structures and activities in the more-than-human world physically, mentally, emotionally, and spiritually. It is not the case that decolonial infrastructures are universally *good* for ecologic relations, and colonial infrastructures are universally *bad* for ecologic relations; rather, it is that decolonial infrastructures make visible some of the taken-for-grantedness of colonial infrastructures. Decolonial infrastructures typically facilitate ways of moving through the world that are continuous with the more-than-human world and that highlight the animacy and intrinsic value of the more-than-human world (Chilisa 2019).

The decolonial view of more-than-human worlds—that is, a common space of existence in which each individual and community crafts distinctives spaces of existence (van Dooren 2019)—is not a utopian view of perfect peace among all beings. That all heterotrophs have to eat other beings to live is, in the words of Cora Diamond (2003), one of the difficulties of reality. Every flourishing "involves a constitutive violence . . . some collectives prosper at the expense of others" (Ginn et al. 2014, 113). Nor should the decolonial view be romanticized or stereotyped and homogenized in a way that does harm to peoples. As Tema Milstein and José Castro-Sotomayor (2020) make clear, all people are equally embedded in the more-than-human world and have their own ecocultural identities. Some ecocultural identities recognize this embeddedness and others do not, but there is no such thing as a lack of embeddedness or an absence of an ecocultural identity.

There are many efforts to decolonize anthropocentric infrastructures. Scientists are turning to Indigenous farming techniques for ways of responding to heat and drought (Gilbert 2021; Settee 2018). Linguists and language teachers are looking to Indigenous languages for ways to move toward a more equitable way of living in the world by transforming (Meighan 2021) and animating (Kimmerer 2013) the English symbol system. Rhetoric and media studies scholars have asked what Indigenous digital interfaces would look like; that is, how would a social network's interface communicate and enable a spirit of sharing and reciprocity, a sense of being part of something bigger (Arola 2017)? Sometimes the search for decolonial infrastructures lead to surprising places. While acknowledging the limitations and problems of Facebook, Kristin Arola (2017) realized that Facebook's interface—not to be confused with Facebook's algorithms and data-collection procedures—afforded much of the communicating-with-her-community that was the goal of the now-folded Indigenous Network social media platform. This is not a claim that Facebook *is* a decolonial space; rather, it speaks to the ability of people to practice decolonial techniques in varieties of settings and with various technologies.

Continuing in the realm of decolonizing the digital, Arola (2018) proposes ethical, relational, and material approaches to digital design that are founded on land-based rhetorics. Land-based rhetorics foreground that experiences in digital spaces are shaped by and assist in shaping the embodied interactions of people in the *topos* of the biosphere. A criterion that Arola provides for a land-based digital design rhetoric is that it acknowledges people's sensate experience and memories of those experiences. In other words, how people encounter digital spaces is understood not as a separate, disembodied realm but as derived from connections made between sensate experience and the memories of the consequences of those experiences (Arola 2018, 211).

Indeed, as Angela Haas (2007) writes, hypertext and multimedia did not even originate in digital spaces; American Indian communities used *wampum* belts and strings to extend and connect memories long before the Internet came into being. To accept a gift of a *wampum* belt or string was also to accept a degree of communal responsibility: *wampum* records are a "living rhetoric" (80) that are kept by rereading the text through communal memory and performance. *Wampum* embody *memoria*-in-*topoi* by putting memory practices into an interconnected, and in this case highly material, commonplace.

Haas's study of *wampum* belts invites a level of Western epistemological and technological humility by showing that the nonlinear designs of memory storage and retrieval methods of the Internet are *fellow* innovations, not *founding* innovations. *Wampum* belts and strings also offer another caution for engaging in decolonial knowledge-making. *Wampum* was appropriated by colonists and used as a form of currency (Haas 2007). Participating in decolonial infrastructuration involves, indeed, necessitates critical, careful, and reflexive methods to navigate the space between engaging Indigenous knowledges and practices and appropriating Indigenous knowledges and practices in the of service of existing anthropocentric infrastructures (Todd 2015). Land acknowledgement statements, for example, range from purely theatrical recitations leading to business-as-usual to profound and only partial forms of engaging with Indigenous peoples in acts of reconciliation (Native Governance Center 2019).

So, what do decolonial infrastructures look like? Shortly, I will provide a closer look at the infrastructuration of "Reclaim the Void" and how it intervenes in a colonial, anthropocentric infrastructure of mineral mining; but for now, I share some insights from Zoe Todd (2015) and Bagele Chilisa (2019). Scholars engaging in decolonial research paradigms—which are themselves part of decolonial infrastructure—take the time and care to acknowledge, for example, that anthropologist Tim Ingold's "ontology of dwelling" stands alongside a vast, varied, and long-held tradition of Indigenous Place Thought (Todd 2015, 246).

Decolonial research paradigms embrace embodied knowledges, ethics of continuity, and a relational ontology—that is, "a reality that takes into account our 'being' relations with the earth, the living, and the nonliving' in an ongoing state of interdependence and interconnectedness" (Chilisa 2019, 99). Within a relational ontology, the more-than-human world is filled with *relatives*, be they other humans, the land, the cosmos, or ideas. The cosmos is the spiritual realm, and "any exercise that increases connection or builds relationships is spiritual" (103). An ethics of continuity asserts that although people are always walking-worlds-into-being, in different place-based cultures and epistemologies, it is imperative to think and act in reference to our relations (Donald 2010). However, decolonial paradigms do not dissolve the individual; care is taken to "construct reality in such a way that the I does not overshadow the other and community" and "the we does not shadow the I" (99). Individuals-in-communities is the basic unit, and *memoria* are embedded in but not determined by *topoi*.

An analysis of two anti-Anthropocene worldbuilding projects, the Pink Chicken Project (2022) and Reclaim the Void (2022), offers insight into *how and to what consequences* (de)colonial infrastructures take shape. Both are digital projects: the Pink Chicken Project exists entirely in digital spaces, and Reclaim the Void lives in both digital space and sacred physical place. As worldbuilding projects, both the Pink Chicken Project and Reclaim the Void disclose the world differently as they engage in activities of transformation. However, while both offer insight into the making of the Anthropocene, I argue that only Reclaim the Void participates in infrastructuration for unmaking the Anthropocene.

(UN)MAKING THE ANTHROPOCENE: THE PINK CHICKEN PROJECT

I do not think it is a coincidence that the Anthropocene has seen a deluge in thinking *with* birds. Birds can be found across the more-than-human world, from the avians inhabiting the poles to the open ocean pelagic seabirds. While dolphins may help people think about what ways-of-life would look like in different media (e.g., the ocean [Jue 2020; Peters 2015]), birds make the medial environment we share—Earth's terrestrial atmosphere—show up differently. Without external technologies, birds can traverse vertically what humans can navigate only horizontally. Birds have (1) inspired rethinking evolutionary processes through consideration of the roles that agency and aesthetic preference might play in sexual selection (Prum 2017), (2) helped peck away at the Great Man Theory—a theory that the major events of history can be attributed primarily to the impact of highly influential men—by suggesting that the compositions of Wolfgang Amadeus Mozart did not come from an otherworldly nowhere but came from a very real, embodied somewhere—including from a life shared with a starling (Haupt 2017), (3) revealed the inadequacies of anthropocentric searches for cognition (de Waal 2016), (4) lived with people in cooperative ways-of-life while retaining profound otherness (Clary-Lemon 2022; Macdonald 2014; Montgomery 2022), (5) invited rethinking the concepts of species and extinction from ideas that were principally concerned with physical morphology to concepts that are embedded in the world-making projects of beings (van Dooren 2018), and (6) illuminated the ethical considerations at stake in the wake of ontologic openness, as in the case of birds imprinting on people (Giraud 2019).

Thus, it is not surprising that much thinking has gone to the birds. What is surprising is the relative lack of attention given to chickens (with some notable exceptions like Josephson [2020] and Potts [2012]). Chickens and chicken bones are visible manifestations of the Anthropocene. Their bodies have changed—from their genetics to their skeletal morphology and bone geochemistry—and the remains of those bodies, Carrington (2016) points out, are a good candidate for the fossils that will define this epoch in geologic record. Yet, despite their abundance, chickens have not influenced more-than-human thinking in the same way that other charismatic avian, like bowerbirds, have.

In terms of number of animals killed, chickens are the world's most consumed animal (in terms of pounds of meat, more pig meat is eaten). Around sixty billion broiler chickens are killed every year. Although chickens have millenia-long history with humans, the scale of chicken farming took a dramatic turn at the end of the Second World War when the United States government partnered with a poultry company to host a competition for the production of the "chicken of tomorrow"—produce chickens that would grow fatter faster. Since then, through a combination of breeding, diet, and farming practices (including indoor, confined feeding operations), broiler chickens have five times the body mass in 2022 than they did in the mid-twentieth century (Bennett et al. 2018). Where chicken meat was once a rare dish for special occasions, it is now a staple food for much of the world. In 1960, an average of 4.4 pounds of chicken meat was consumed per capita in the United States; now, around 98.7 pounds per capita are consumed—a 2,143 percent increase in just over sixty years (National Chicken Council 2021).

One project that does attend to chickens, and in a highly provocative manner, is the Pink Chicken Project. The Pink Chicken Project (PCP) is a website that offers a "speculative stirring" (The Pink Chicken Project 2022). The PCP urges intervention in chicken's DNA to turn their bones pink, thus creating a pink layer of chicken bones to mark the stratum of the Anthropocene in the geologic record. The PCP states that all living chickens could soon have pink bones due to the proliferation of gene editing technologies, the low cost of these technologies, and lack of regulation combined with the housing of billions of chickens in relatively few locations. In addition to changing the color of the bones, the PCP would encode a manifesto in the chickens' DNA that would mark with horror the environmental and social injustices being perpetuated by oppressive systems. The PCP is highly paradoxical—suggesting the use of nonconsensual modification to highlight how humans have engaged in nonconsensual modification and attempting not to de-occupy but to re-occupy the geologic stratum with different colored bones. The website notes these paradoxes and refuses to resolve them, claiming that the paradoxes "allow complex entanglements to remain complex" (The Pink Chicken Project 2022). The PCP functions as something like a Swiftian "A Modest Proposal" for the Anthropocene in that it is a satirical gesture meant to raise awareness of issues through provocation.

Stratum-Altering Rhetorics: Colonial Infrastructuration

The PCP is a provocative proposal meant to spark conversation around globalization, industrial agriculture, social justice, and technological systems of domination. Here, I argue that the stratum-altering rhetorics of the PCP are both critical of and complicit in maintaining the colonial infrastructures of the

Anthropocene through the (hypothetical) project's dependency on the infrastructures it critiques and through its lack of attention to ways of unmaking the Anthropocene. As I mentioned previously, the PCP embraces the complexities of their paradoxical proposal. Thus, my analysis here works with the PCP in further illuminating the complexities at play in such a proposal.

The many paradoxes of the PCP begin with the acknowledgment that the shared commonplace of Earth, the *topos* in which all acts of worlding take place, is in danger of becoming uninhabitable (The Pink Chicken Project 2022). Yet, the PCP aims to leave traces in the strata and messages in DNA that will be available "for geologists of the future" (The Pink Chicken Project 2022). What geologists of what future if the planet is uninhabitable? As rhetoric, stratum-alteration is both too far-sighted to invoke change in living-well-with-others in the here and now *and* is a perpetuation of colonial infrastructures of expansion. Rather than the Manifest Destiny of westward spatial expansion, stratum-alteration relies on a temporal expansion. The same infrastructures that enable gene editing and deciphering today will still need to exist in this imagined future—and this is a Schrodinger's future in which life both exists and does not exist—to be legible. Encoded messages rely on particular infrastructures and literacies to be decoded. The PCP depends on the infrastructures it critiques to have any meaning; without those infrastructures, the PCP is just a layer of pink detritus. The layer becomes a pink *topos* without *memoria*, a commonplace without individuals.

As stratum-altering rhetorics, the PCP fights instrumentalization of the more-than-human world with a more colorful instrumentalization of the more-than-human world. In the PCP, chickens are disclosed as mere means to *illuminate* the life-threatening infrastructures (e.g., economics, politics, industrialized agriculture, mobilities, and—not least of all—ideologies) of the Anthropocene via remaking chickens through gene editing. As the PCP notes, to accomplish such alteration of chickens' bodies on a global scale requires that chickens be housed in industrial settings. Thus, these stratum-altering rhetorics perpetuate a commodity-crisis paradigm (described in chapter 1), even as the PCP strives to critique commoditized relations with the more-than-human world. There is no *memoria* in the PCP—no individual chickens with biographies of their own; chickens are only *topoi*, a commonplace in which humans can inscribe their own anxieties.

There is a trickier issue here in the stratum-altering rhetorics of the PCP that deserves some unpacking. It is the matter of how and to what extent it is ethical to use technologies to alter the more-than-human world. What if, instead of using gene editing to color chicken bones and encode a manifesto, chickens were edited to be extremely slovenly creatures uninterested in movement and to be incapable of feeling pain? What if their genes were edited in a way to answer Bentham's (2007) question of "can they suffer?"

(9) with a scientifically confident, "No. They cannot"—such as in lab-grown, synthetic meat? However, a more nuanced response is called for, not only to analyze satirical projects like the PCP but to think through the sincerely proposed uses of gene-drive technology, like using the technology to make extinct mosquito species that spread malaria to millions of people every year by editing genes to sterilize females (Saey 2022).

For some, the response to the hypothetical painless chicken or the purposeful extinction of a species—even a species that parasitizes and causes suffering and death of many humans every year—will be a visceral one that this is wrong; it is "messing with nature." For others, the painless chicken and sterile female mosquitoes may sound like an ideal solution: no suffering, no ethical dilemma. Both strong views miss some important nuance to which I will attend, and in doing so, I will also better highlight the importance of infrastructuration as a concept and how it relates to more-than-human ethics of care.

The view that any technological intervention in the more-than-human world is "messing with nature" is, itself, reinforcing a nature/culture or nature/technology divide. The human world does not exist apart from the more-than-human world; there is no possibility of intervening from the *outside*. There is no stable, Platonic ideal of nature that humans can "mess up." Humans may be the most efficient but are not the only colonizing species who have the capacity to radically transform ecosystems. Kudzu may have been brought to the United States by humans, but the agency of kudzu-in-southeastern-United-States-climate to transform landscape and ways-of-life exceeded human imaginary capacity for control. The more-than-human world is abundant with capacities for transforming and being transformed by others. Indeed, biodiversity would not exist without a dynamic biosphere.

The point here is not to excuse human activities, like the editing the PCP proposes, under some justification that "intervention is natural." The point is to recognize that there is no such thing as *intervention* when understood from an outside perspective; there is only complex, entangled relations of becoming or, in the ethos of Karen Barad (2007), *intravention.*

Clearly, there are actions humans take on and for the more-than-human world that most people find responsible and good. My companion shepherd mix is now a senior dog and lives with arthritis that has added pain to her ambulation, and this is a dog who loves not to walk but to bound places. I doubt many people would object to the, inevitably, nonconsented injections of polysulfated glycosaminoglycan that my partner and I administer to help ease her pain. I will get at why I believe this is a less-objectionable intervention in a moment, but first, I must attend to the other view aforementioned about suffering and ethics.

The view that suffering is the only rubric from which to assess ethics comes from a highly constrained ethical imagination. Rendering a nonhuman being incapable of suffering may solve the issue of vulnerability to pain, but it does not address the beings' right to flourish in their own ways-of-life. Extinction is not only an end to a particular morphological configuration (e.g., a being that is physically recognized as a chicken) but also a process that ends ways-of-life (van Dooren 2018). A genetically engineered, pain-immune, lethargic chicken is *not* making chickens invulnerable to suffering; it is unmaking chickens—extincting a chicken way-of-life. Many chickens' ways-of-life are already extinct in factory farms in which they are confined and unable to scratch, sun their wings, cluck to their incubating eggs. Thus, the genetically altered chicken who cannot suffer may address the ethical responsibility to minimize pain, but it is insufficient in a responsibility to allow for flourishment.

The other issue in accepting that a chicken without nociception (i.e., the ability to detect painful stimuli) would resolve ethical dilemmas is that it reduces ethics to a simple checklist. Precisely because the more-than-human world is always in a dynamic process of becoming, a static checklist of ethics is fundamentally *not* doing ethics. When it comes to questions of ethics, "There can be no generalizable or final answers . . . only shifting, specific, and always relational possibilities for better and worse forms of being" (van Dooren 2019, 9). And while the "complexity of real worlds and the unavoidable partiality of our understanding cannot be an excuse for indifference and inaction" (12), those actions cannot be dictated by an inert, *a priori* checklist. More-than-human ethics of care are attentive to particulars and to consequences, to the way (in)actions impact others. A technological intravention that would attempt to sideline ongoing and messy engagement is a hallmark of the technological hubris and attitudes of domination characteristic of colonial infrastructure. Ethics cannot be outsourced to technology.

Let me move away from the painless-chicken thought experiment for a moment to connect the PCP with a physical, materialized project. Jennifer Clary-Lemon (2022) traces the ways-of-life of barn swallows, an endangered species recognized by the Canadian government's Species at Risk Act. The three major factors contributing to their decline are the effects of anthropogenic climate change on their breeding grounds, the large-scale declines in insect populations, and the loss of habitat as farming methods have become more industrialized. As a response, the government has offered small, wooden structures—"replacement habitats"—for swallows in which swallows can nest. The problem is the swallows do not use these boxes. The boxes primarily function on an understanding of nest-as-place (i.e., that providing an alternative space to live is sufficient) and not nest-as-process, one that takes into account the ways-of-life of barn swallows and how they settle in breeding pairs in groups of up to one hundred thousand birds and make labor-intensive flights to gather mud for nest-building.

In contrast to the rather anthropocentrically constructed replacement habitats (which are even referred to as "residences"), Clary-Lemon describes the acts of care of Ray Lammens at Spearit Farms. Lammens has a large barn in which he processes and packages his produce. It also happens to be the nesting site for thirty to forty breeding pairs of swallows. The problem is that the presence of the birds poses a health and safety risk due to the proximity of bird excrement to food. Many farmers choose to exterminate the swallows, but Lammens chose to have a 70′ × 50′ tarp created to provide a ceiling to separate the swallows and the produce. He also provided a fan to keep the tarped-ceiling from overheating and a shallow pond for access to mud. Lammens is enacting a highly technical intravention, but one of caring for human health and safety and of careful attention to enabling the swallows' ways-of-life. Lammens uses technology in an ongoing ethical engagement with swallows.

The trip through a philosophical thought experiment of painless, stationary chickens and barn-swallow intraventions may have seemed like a detour, but it provides the grounds to return to the PCP and assess a final—but by no means *the final*—paradox of the project beyond a kneejerk technophobia or technophilic response. The PCP articulates its environmental concerns of climate crisis and waste-world-making with social justice issues of racism, sexism, ableism, homophobia, and transphobia, *yet* it participates in infrastructuration that perpetuates those oppressions rather than the infrastructuration offered by those being oppressed. The PCP perpetuates colonial infrastructures by offering no possibilities for the *otherwise.*

Otherwise is a key term in feminist, queer, disability, and Black vegan scholarship that considers the word "a concept, an analytics, a method, and an ethico-onto-political commitment to the insistent of the possible against the pull of the probable" (Meek and Fontanilla 2022). For example, rather than making *accommodations*—a reactive process—in the classroom for neurodivergence, there is advocacy for seeing the classroom *otherwise* and striving for proactively employing *accessible* material arrangements and pedagogical practices (University of Minnesota 2022). The move from accommodations to accessibility is not a mere substitution of words. Accommodating and making accessible differently orient people to the task of teaching. *Otherwise* is a worldbuilding term of possibility and is therefore profoundly rhetorical, as Aristotle (1991) notes that the realm of rhetoric is neither the inevitable nor the impossible but in the realm of potentialities. The PCP articulates its cause with the oppression of people but does not engage with the strategies of resistance of seeing the world otherwise. How are humans to live with one another? How are humans to live with chickens? How are humans to live with the more-than-human world? Unlike Lammens's farm, the PCP does not address these questions. Instead, it offers an intervention that perpetuates the harmful conditions of what is, and only in the manifesto embedded in the DNA found in the decaying chickens' bones does it offer a hope that *others* have found ways of living differently.

Thus, the stratum-altering rhetorics of the PCP may be an appropriately dramatic response in terms of imagining a globally scaled rhetoric to illuminate a globally scaled problem. However, it is an ethically unsuitable *topos* for shortening and thinning the Anthropocene (Clary-Lemon 2019, 22). The PCP helps stabilize anthropocentric infrastructure rather than inventing infrastructure that affords new forms of. The PCP erases the chickens on which the project writes, adds no new doors for adjacent possibilities of living otherwise, and is itself infrastructural to the infrastructures it critiques. The question then becomes, *Is it possible to have globally scaled rhetorics that attend to individuals-in-communities, open possibilities of living otherwise, and offer alternative infrastructure?* The short answer is *yes*. The long answer is the next case study.

(UN)MAKING THE ANTHROPOCENE: RECLAIM THE VOID

While there are many social and ecological reconciliation projects taking place around the globe, the remainder of this chapter focuses on a particular project originating in Western Australia. To help contextualize this project, I first need to provide a broad sketch of the continent's colonial history.

In 1770, Lieutenant James Cook claimed possession of the east coast of Australia for the British crown. The land was deemed advantageous for multiple reasons; as a penal colony, Australia would serve to simultaneously alleviate the burden of overcrowding in British prisons and would expand the British Empire with a significant southern colony. Eighteen years later, eleven British ships—carrying 1,500 British convicts, marines, and civilians—arrived at what is now called New South Wales, an area that consists of tribal lands for multiple Indigenous Australian communities. In the ten years that followed, it is estimated that the population of First Nations people decreased by about 90 percent due to dispossession of land, which fundamentally changed ways-of-life; disease, which included smallpox, measles, influenza, and the introduction of venereal diseases; and direct conflict, which led to the deaths of some 20,000 Indigenous peoples and approximately 2,500

European colonizers (Australians Together 2022). Western Australia was colonized in 1829, when Captain Fremantle claimed the land as the Colony of Western Australia (Swan River Colony) in the name of the king of England on May 2 (Find and Connect 2022).

Today, the state of Western Australia is considered one of the "most attractive" regions for mining in the world (Mining[Dot]Com 2022). More than fifty minerals are mined in Western Australia in about 125 operating projects (The Chamber of Minerals and Energy in Western Australia 2022). Western Australia is also home to Ngalia peoples, and mining activities leave holes that have "scarred" sacred land (Reclaim the Void 2022). Reclaim the Void (RtV) is a collaborative project of restoration.

On the project's main page, RtV is described as an idea born from the grief of Ngalia elders over witnessing the harms done to the land by industrial mining practices. The mining holes are considered *scars* on a living country. Here, I quote at length—with permission from RtV's website—to preserve the project's purpose in the community's words:

> Reclaim the Void was born from Ngalia elders in Leonora, Western Australia, expressing their pain and grief at "those gaping mining holes left all over our country." The idea was conceived to symbolically "seal" one of the holes with an artwork expressing the story of country.
>
> Country is alive with story, song, dance, law, Tjukurrpa. When we wound country, we wound ourselves, and end up with a scarred physical and cultural landscape. This project carries the desire for healing country, healing community, and healing ourselves. It is about acknowledging the hurt and contributing to restoration. It offers people the chance to learn about country and culture.
>
> The vision is to cover a mining pit with a large-scale "dot" artwork made up of thousands of handmade circular rag-rugs woven from discarded fabric. Woven by people from all walks of life and backgrounds, the rugs will be joined together into a giant textile artwork which shows an overall pattern that carries the story of the Tjukurrpa of the country on which the pit is situated.
>
> Reclaim the Void is a bold cross-cultural project. It seeks to raise awareness of the story of country and its importance in Aboriginal culture in both its physical and spiritual dimensions. We invite you to join us. (Reclaim the Void 2022)

Here, RtV gestures at the continuity of grief and celebration, at the ways in which destructive practices—like industrial mining—are also embedded in relations and networks that afford reclamation. The harm to country is not undone by the collective tapestry; rather, the tapestry is materializing a different, affirmative way of being a global community. Again, to quote from the project, "This project isn't 'anti-mining,' it's about walking forwards together in an act of healing, caring for country and community engagement."

Reclamation Rhetorics: Decolonial Infrastructuration

RtV participates in oppositional and affirmative infrastructuration through the project's aesthetic, techniques, and ethics of reclamation and to show how these aspects of reclamation contribute to thinking on more-than-human ethics of care. RtV participates in decolonial infrastructuration that is simultaneously oppositional to colonial infrastructures of the Anthropocene while being affirmative of *anthropos* by offering collaborative and community-building ways of being human. Crucially, RtV envisions and enacts possibilities for living *otherwise*. RtV engages aesthetics of reclamation to heal scarred land and promote social justice. Reclaiming country is reclaiming the *anthropos*.

> The RtV team estimates that the project will need about 5,000 handwoven rugs to join together to cover one of the mining holes. Each rug is an art project for an individual, and each rug becomes a "dot" in the larger design. The tentative design features vibrant colors, circular patterns, and lines of movement. The design is tentative because "the final story-artwork will be devised according to cultural principles" and in relation to rugs received (Reclaim the Void 2022).

Earlier, I addressed the decolonial perspective that the world is filled with relatives. The idea and design of RtV is *familiar*—known to and abounding with relations—to Kado Muir (a Ngalia custodian and cofounder of the Ngalia Heritage Research Council) and Ngalia peoples. The artwork is one of Muir's mother's paintings, who was a Ngalia leader who painted to preserve stories of country and community. For Muir, the inspiration for RtV connects with Indigenous wisdom beyond Western Australia. He finds resonance in Louis Riel's, a Métis leader and one of the founders of the Canadian province of Manitoba, belief that it would be arts that bring peoples' spirits back.

It is not only Indigenous wisdoms that are woven into this tapestry. In the written instructions RtV provides for rag-rug weaving, it is noted that the weaving technique being used—"weft twining"—originated in Scandinavia and made its way to the United States in the 1800s, and that RtV has adapted the technique by drawing from several sources. "Weaving and twining are as old as the human story itself. . . . Reclaim the Void honours all traditional and contemporary weaving practices and the weavers and twiners who keep the fabric of the world strong" (Reclaim the Void 2022).

The joined rugs do not only *cover* the scars left from mines. The rugs are sociopolitical and aesthetic acts of social reclamation. Sociopolitically, RtV has partnered with—as of February 2023—over thirty schools from pre-K to high school to engage students in the process of weaving rugs for the tapestry. Students are involved in learning the techniques and contributing to a project born from the Ngalia community. Importantly, the school partnerships afford students the opportunity to learn from and support community-directed initiatives. Further, the school partnerships are partially supported by Lotterywest and the Government of Western Australia's Department of Local Government, Sport and Cultural Industries, whose financial support has allowed schools to sponsor week-long artist residencies to engage students in weaving rugs for the tapestry. RtV has over a dozen supporting partners, including several governmental institutions.

A quick and easy critique of government's financial support is to say that these governing bodies are pacifying a community from which they reap substantial financial benefits by allowing the very extractive mining that the community is attempting to reclaim. That critique is not *wrong*—mining is a substantial source of income—but it is incomplete. RtV is a project "informed by Ngalia Western Desert traditional knowledge, spirituality and culture and is immersed in the cultural custodianship of elders past, present and emerging" (Reclaim the Void 2022) and the entanglement of government dollars into a Ngalia-born project is, at least, a material recognition of the Ngalia community as part of the larger Western Australia community. *Recognition* does not mean that RtV is receiving equitable or even equal contributions from the government, but the financial contributions do amplify the *voice* of RtV via more platforms and economic support.

Aesthetically, RtV has welcomed all to share their rag rugs and their stories and be part of a community of healing; the project's website attends to accessibility by sharing rug-weaving techniques with written and video instructions, including how to prepare a loom. Unlike the PCP, RtV does not depend on existing anthropocentric infrastructure for legibility. Art works in a different register. As Whitson and Poulakos (1993) write, aesthetic rhetorics attend to the senses, and beings have art in order to live. "[W]hile the epistemic relies on the cognitive mechanisms of induction and deduction, the aesthetic relies on the sensual process of seduction" (142).

The potentiality of art to impact through seduction relates to what Abigail Feldman (2021) calls "aesthetic of access" (7). An aesthetic of access is not a singular mode or medium, but an attention to affordances of media, the values of peoples, and the possibilities for continued relation. Aesthetics are, to borrow from Kenneth Burke's observation on music, "sheer rhetorical force, unencumbered by information" and this is "an enlivening force—sheer energy—with a unique capacity to mingle with and transform bodily energies" (Hawhee 2009, 27–28). Such an enlivening force can be seen throughout the more-than-human world with the magnificent display of patterns on nonhuman animals and the brilliant colors of flowers and fungi. Aesthetics are thus more-than-human rhetorics.

"An effective art of the Anthropocene is Indigenizing the Anthropocene . . . one that directly engages with the structural violences of heteropatriarchy and white supremacy as they shape discourse and praxis. This is where the work of Indigenous scholars and artists promises to speak back, reshape, and change the direction" (Todd 2015, 248). The rag rug tapestry speaks back, reshapes, and changes the direction of the Anthropocene. Instead of extracting from and abandoning the land, RtV offers a prolonged engagement of attending to and weaving for the land.[1] Where Eurocentric perspectives see deserts as "wastelands"—a waste of land—from which it is acceptable to exploit resources (Rosenfeld 2019, 9), the Indigenous perspectives in RtV understand the land as a relation worthy of care. The individual and joined rugs and the commitments that afford their materializing in the world are decolonial infrastructuration.

Weaving is a technique of reclamation. In the written instructions, RtV notes, "Don't worry if you aren't 100 percent sure about what you're doing, or make a mistake. Everyone's in the same boat! Just keep going and your rug will turn out beautifully. . . . It's absolutely fine if yours are slightly different." Weaving is a technique that resists the structural violences that shape the Anthropocene. As Marcel O'Gorman (2019) writes, to weave is to accept the possibility that ideas may not materialize according to hopes and expectations; weaving embraces "the unknowability that comes with haptics." Weaving is a way of making that is part of decolonial infrastructuration, like *gambiarra* (Fonseca 2017). *Gambiarra* is a term used in Brazil to describe improvised fixes to everyday problems. *Gambiarra* usually denotes informal repairing—often necessary because of a lack of access to materials or techniques, but more importantly, it resists a paradigm in which there is a singular *right* way to innovate, tend, or repair. Both *gambiarra* and weaving draw attention to the difference between technologies and techniques. The Internet, for example, may be afforded by screens, wires, energy, and an exponential proliferation of binary code, but there are many ways of interacting within the spaces those technologies make possible. Some ways limit potentials for flourishment; others expand it.

Felipe Schmidt Fonseca (2017) describes the spirit of *gambiarra* as "distributed, *in situ* creative action" that establishes closer connection to the materiality of the world. *Gambiarra* resists dehumanization by allowing people to acquire agency through tinkering. The *resistance* of *gambiarra,* and of weaving, to dehumanization is important to decolonial infrastructuration. What is being discussed throughout this book is learning to *live* well with others.

Within colonial infrastructures, one proposed solution to the problems of the Anthropocene is for humans to voluntarily extinct themselves. I understand the deep ecological grief that can invite such solutions—a grief that discloses humanity as a parasite (Pauly 2014) of Earth. Voluntary extinction seems to be a radical solution; what could be more extreme than placing ourselves in the red on International Union for Conservation of Nature list? However, on closer inspection, it is not as radical as first appears.

Voluntary extinction still imagines humans as in control; humans caused the Anthropocene, and humans can repair things by extinguishing themselves. It ignores that conditions of the biosphere have already been altered, and so extinction would not so much solve the problem as potentially abate the problem and just remove humans from having to witness whatever unfolds from the changing conditions. More importantly, as already discussed in this chapter, not all humans caused the Anthropocene, and RtV offers techniques for living otherwise.

The techniques for living otherwise get at the project's ethics of reclamation, which include the enactment of difference-in-solidarity and ethical consideration of individuals-in-communities. Beginning with the ethical enactment of difference-in-solidarity, RtV welcomes all peoples to weave their own rugs and tell their own stories to help heal lands and create new stories. There is no litmus test for being a rag-rug maker, no required level of weaving proficiency. One rug is not better, nor does it tell a more meaningful story. There are only different people, different rugs, different stories.

However, difference alone does not make for solidarity. So, the RtV requests that each individual rug needs to be kept within a narrow color spectrum. From a distance, each rag rug should appear to be of one color, so that the rugs may then be combined in ways to be part of this textile mosaic. In the instructions, RtV affirms the beauty of highly variated and colorfully patterned rag rugs *and* says that those rugs are not suited for *this* project. Certain exclusions are necessary to make certain relations possible (Giraud 2019, 173). The balance in RtV—between making one's own rug to one's ability and doing it *within* the needs of the community—is the decolonial work of constructing reality in a way that the I does not overshadow the community and the community does not overshadow the I (Chilisa 2019, 99).

The I-we—the individuals-in-communities—negotiation is attended to in all aspects of RtV, which is another aspect of RtV's ethics of reclamation. In addition to the materiality of the joined rugs, RtV wants to create a "rugalogue," in which they gather and preserve the stories of the rug makers. Both the emerging material and virtual rug(alogue)s are enactments of *memoria*-in-*topoi*. Individual rugs are materials woven with their own *memoria*, and they join to form a *topos* that tells a story of community and country. RtV is a weaving of individuals-in-communities, *memoria*-in-*topoi*, differences-in-solidarity; in other words, RtV is a material and visible manifestation of a relational ontology and ethics of care and interconnection.

Further, RtV's practice of gathering and preserving stories in a "rugalogue" connects and resonates with other Ngalia practices of using the affordances for preservation and sharing. The Ngalia language is listed as critically endangered, with only the Muir brothers (including Kado) speaking the language fluently. Kado Muir has participated in the 50 Words Project, a repository that holds recordings of fifty words spoken in Australia's Indigenous languages and maps those languages onto the land from which they came into being (50 Words Project 2022). Other acts of Ngalia preservation using digital technologies include: the use of GPS devices to document coordinates of sacred sites, the creation of videos showing activities of Ngalia ways-of-life, and the creation of textual and visual Ngalia dictionaries. Technologies like the Internet provide deep reservoirs of anger, resentment, and anxiety, but that is not all these technologies can be. They *can be* "technologies of recovery" (Gallon 2016). RtV is thus part of decolonial infrastructuration that discloses more-than-human world(s), including the more-than-human digital world(s), differently.

Globalization shows up differently in RtV. Digital technologies can be used not only to produce energy-intensive cryptocurrencies and circulate democracy-destabilizing conspiracy theories. Digital technologies also afford techniques of ethical and aesthetic engagement. In RtV, the *anthropos* of the *Anthropocene* also show up differently. Humans are not just extractive but also creative forces on Earth, not just consumers but also custodians, and are not just involved in waste-world-making but also capable of aesthetic-world-repairing. RtV is not going to halt the Anthropocene, but it intravenes in the Anthropocene in decolonial ways. RtV contributes aesthetics, ethics, and techniques that are infrastructural to living well with others.

RECLAIMING THE *ANTHROPOS* IN THE IRREPARABLE

The affirmative analysis above of RtV's global enactment of an aesthetic of inclusive invitation should not be read as an assertation that the project rests on some onto-ethico-epistemological pure ground, making a clean break from the anthropocentrism of the rest of the world. The material body of the project uses human textiles to create rugs using human techniques. What happens to these rugs as they weather? Will strips of fabric be swept to the ocean and carried by currents into some pelagic debris assemblage? Perhaps Western Bowerbirds, known to decorate primarily with bones and shells, will come to incorporate fabric scraps in their displays. The long-term relations of this tapestry are unknown.

While the long-term relations remain in the plane of possibility, the immediate relations are rife with constraints. The collection of rugs into a mosaic tapestry does not, literally, heal the land in the sense that it cannot restore what was mined, cannot return the land to its pre-extracted state. The giant mining holes are, mostly, irreparable—and irreparable situations create an exigence for precaution rather than reaction (Cox 1982). The process of global inclusion—soliciting rugs woven by an individual and connected to a biography—enacts techniques that work against the numerical reduction of individual lives perpetuated by Big Data, but those techniques are still, necessarily, embedded in a commercial and military infrastructure of digital communication technology. Further, the size of the rug—necessarily large to cover the hole—calls for a drone-eye's view to be able to grasp the design in its entirety. The *view from above* has invited humility from some, such as astronauts describing the experience of seeing Earth from space, and hubris from others, such as a talk I heard exhorting humans to think of and act on climate change by taking a god's eye view of Earth. Whether inviting humility or hubris, the *view from above* is often achieved through technologies enabled by anthropocentric infrastructure. Like the Pink Chicken Project, RtV is still dependent on and working within some problematic infrastructures.

There is no exit, only staying with the trouble. RtV is not a *solution* to a problem; the project is a response to a situation. RtV cannot restore the land, only disclose the land differently, invite transformation, and create the possibility for new lines of affinity. RtV is in important respects a microcosm of the global work of trying to live well in the age of the Anthropocene. Many of the consequences of the Anthropocene are irreparable. Species have been driven to extinction. This cannot be undone. Even with technoscientific de-extinction projects, whatever species emerges from the process is not *the same*. The species may be genetically and morphologically representative of the extinct species, but that is not all a species is. The species was also about a way of living in particular environments. The very idea of *restoration*, always a bit suspect from a relational ontology, is absolutely untenable in an age of rapid technological and ecological change.

Healing is less about *restoration* than *reclamation*. In a relational ontology, *restoration* is a logic-defying concept. The RtV tapestry does not *restore* the land, so much as it works to reclaim it by inviting a global community into an aesthetic and ethic of care. Here, reclamation is an invitation to live otherwise. The tapestry ravels and unravels, has ways of flourishing and inviting flourishment, and also has ways of being vulnerable and inviting vulnerability. The Anthropocene is a *doxastic* age: It does not afford easy, assured solutions to its crises, only (more-than-)human responses to the situations that arise. The lack of assurance makes the work of living well exceedingly difficult and places the (more-than-)human world in a state of great precarity and vulnerability. But it is also the *doxastic* nature of the era—characterized by ambiguity and becoming—that offers hope of flourishment, for it is in that ambiguous and continually unfolding space that the *otherwise* exists. Humans are still disclosing themselves—still *becoming human*—and thus what it means to enact *humanity* remains a fluid concept, one subject to transformation. Humans cannot unmake the voids we have created, but we can begin to build ethics, politics, energies, discourses, values, and ways-of-being that enable us to live otherwise. RtV is a door to adjacent possibilities.

A READINESS TO BUILD MORE-THAN-HUMAN WORLDS IN DIGITAL SPACES

This chapter addressed how two digital, worldbuilding projects contend with the causes and consequences of the Anthropocene. By imagining a world lined with the remains of visibly genetically modified chickens, the Pink Chicken Project offers a dramatic visualization and scorching critique of colonial mentality and materiality. By their own accounting, the PCP neither attempts nor achieves an emancipatory engagement with colonialism; rather, the project amplifies colonial ontology and power relations through its stratum-altering proposal. Another project, Reclaim the Void, enacts a readiness to build worlds *otherwise* through the decolonial reclamation of land that invites all humans into this community-care practice. The hole will still exist, but the wound will no longer be an open one.

The concept of infrastructuration and how it was elucidated through the two case studies make a few key contributions to ethics of care and rhetorical theories. Infrastructuration—as the dynamic event of worldbuilding with embodied, material, digital, and symbolic elements—provokes important rhetorical and ethical questions.

Rhetorically, infrastructuration animates questions like, *what commitments (e.g., ideologies, lines of affinity) and capacities (e.g., access to what technologies, what techniques of engagement) afford the circulation of X discourses, bodies, artifacts, and ideas? If X discourses, bodies, artifacts, and ideas were to settle and become (impermanently) sedimented, what kind of infrastructure do they build; what kinds of discourses, bodies, artifacts, and ideas does this make likely?*

Ethically, infrastructuration provokes lines of inquiry like, *what kinds of relations (ways-of-relating) afforded X arrangement of embodied, material, digital, and symbolic elements? What kinds of relations (ways-of-relating) are probable and what kinds of relations are possible if X arrangement of embodied, material, digital, symbolic elements were to become settled and (impermanently) stabilized?*

Beyond the questions that this engagement with infrastructuration provokes, this chapter also expands thinking on *memoria* and *topoi* for the worldbuilding potentialities of rhetoric. *Memoria*, once a canon of rhetoric, has received less attention as technologies for storing memories outside of the body have proliferated. However, *memoria* is not only about the storage of facts or the recitation of lines. Memories get inscribed in bodies, even when their details escape symbolic recollection. Were the PCP a reality, the chickens' bodies would remember the alteration to their code, whether or not anyone was around to decode the manifesto that explained why. Legibility of memories may wax and wane, but impact remains. This is why it matters what memories are made in the *topos* of the biosphere. This is why RtV is an important intravention. In addition to the inscription of scars, RtV is an attempt to build a commonplace of community in the commonplace of Earth.

This chapter also offered the feminist, queer, and intersectional notion of the *otherwise* as an important concept for (more-than-)human ethics of care, particularly in the Anthropocene. The main element lacking in the Pink Chicken Project, I argued, was any suggestion of how to live in any way other than as exploitive, dominating agents. The PCP *declaims* the Anthropocene in dramatic and provocative fashion. RtV could have functioned similarly, perhaps proposing a similarly satirical project that would scar the sacred places of the commercial miners, leaving holes in the homelands of others. Such a proposal would highlight the environmental racism of industrial extraction and waste allocation and would operate not unlike the PCP by showing how the *whole world* can become the *hole world* if people just keep extracting and moving on. Rhetorically, such a strategy shows the problems of the present infrastructure through amplification, but ethically, it does not offer ways of living together. It is possible to denounce exploitive actions—like referring to the mining holes as wounds—*and* announce ways of living otherwise—like being invited to a communal project of reclamation.

In 2010, Bowker and colleagues wrote that the vast digital spaces and globalized networks afforded by robust information infrastructure required new looks at what *homesteading* means in this new landscape. RtV offers ways of homesteading in digital spaces and in physical spaces where a land ethic really means the entire planet. RtV invites people around the globe to reclaim the land and reclaim the *anthropos* by weaving and sharing to restore earth together. Through the grief over harmed country, Ngalia elders disclose other possibilities for human-land relations. Through the imaginative art project of communal weaving, RtV transforms an aesthetic of extraction into an aesthetic of care. Through the inclusive invitation to participate, RtV engages in a decolonial infrastructuration of entanglement and interdependence.

The next and final chapter of this book aims to continue the work of RtV by offering some practical suggestions for living well with others in digital pedagogy and practice.

STORY TIME: INVENTING A FLOCK, PART 2

"There you are! What are you doing here?" I said to the blue Australorp hen who was sitting in a corner of the goats' barn and whom I thought had been taken by a hawk a few weeks ago. I should have known, given the previous year's experience with the black Australorp, what to expect. I bent down to pick her up and take her to the safety of the chicken coop, and as I lifted her off the ground, I heard the sound of gentle thuds onto the straw. Under the hen were three fluffy chicks. Oh, my. I was immediately in awe and overwhelmed.

Prior to their arrival, the resident chicken flock consisted of nineteen chickens: three roosters and sixteen hens. This was a precarious balance. Depending on the roosters, this ratio *may* work. When the rooster to hen ratio is not right, the risk of rooster aggression is high: rooster-to-rooster aggression, rooster-to-human aggression, and more aggressive rooster-hen mating practices that can leave hens with featherless patches and injuries. It was clear that 3:16 ratio was not quite working out for this flock, as I had already had to put protective cloth skirts over my hens when I started seeing bare spots on their backs. The three new arrivals meant something between a 6:16 and 3:19 ratio of roosters to hens, and only the 3:19 was a sustainable option for the wellbeing of the hens.

For the next several weeks, these were Schrödinger chicks for my partner and me. As the weeks passed and their tail feathers became droopier, we feared they were cockerels (juvenile roosters). If, indeed, these were three young males, how could I ensure the best life for them while also keeping my hens safe?

The morning came when not one, not two, but all three juveniles croaked their first cockerel calls. I took pleasure in hearing them try out the new sounds that their bodies did not quite yet know how to produce, hearing the *becoming rooster*. And my heart sank. These three beautiful birds, who were gentle birds and had only ever known this small area of North Carolina's Piedmont, could not stay here.

We had an infrastructure problem. Our current arrangement of coops and rooster-to-hen ratio did not afford the cockerels living with us. If the cockerels stayed here, we had to (1) keep them confined to a coop with a small run, which did not seem like the flourishing chicken life we were committed to, (2) accept that there would likely be some violent rooster altercations and that the hens would be subjected to aggressive and harmful mating practices, which also did not seem like the flourishing chicken life to which we were committed, or (3) build another coop and add substantially more hens to the flock, which ran the risk of an ever unfolding flock expansion and that, too, seemed outside the conditions of flourishment. Embodying Giraud's (2019) articulation of an ethics of exclusion, they would have to be excluded from *this* arrangement for the flock to flourish.

But *how* do I exclude these three cockerels? They had as much claim to their own flourishment and to my care in supporting that flourishment as any other chicken on the farm. I reached out to farm rescues I knew, but they were chock full of roosters that they were working to rehome. During my most panicked moments, my thoughts went to some very utilitarian solutions. I may eat a plant-based diet, but others do not, and I thought about having a kosher butcher kill them and then donating them to a local soup kitchen. That action would help hungry people and the remaining 19 birds. This was not merely a quick thought that went through my head. We began researching how to enact this plan.

The plan was unsatisfying. Every time I saw the boys scratching and pecking in the yard, I became freshly agitated at the difficulty I was having figuring out how to achieve good lives for them and the rest of the flock and also startled at how readily their planned deaths seemed the most viable choice to me. The infrastructure for killing a backyard chicken—from the bounty of online resources on how to kill a chicken to the ease of purchasing instruments that can be delivered to one's doorstep—is well-worn and easy to travel. The infrastructure for rehoming a chicken, however, is always under construction, requiring more time and effort to navigate.

A friend of mine from animal services told me that she knew of a rooster rescue about an hour away that might be able to take in the stray birds. I located the person who ran the rescue on Facebook and sent a private message. He responded quickly that he was willing to take the cockerels and asked if I could bring them. His address could not be found on any of the GPS apps I used, and the closest I could get was an address of a gas station I supposed was nearby. I got the three young roos in the car and headed off.

For the hour and half drive, I talked—addressing the cockerels—about what was happening. I was relieved that I had a friend who knew this person but also weary of driving alone to an unknown area, to a location I could not quite disclose to my partner because I could not quite locate it myself.

I arrived at the address and was greeted in the driveway by a man ready to meet the new roos. He apologized profusely. He had sustained a stroke and found texting challenging. He remarked how beautiful the cockerels were and said his wife would be home soon to meet them, too. There was a holding coop set up for them, so that they could make sure no new diseases were being introduced to their flock and to give all the birds some time to adjust to each other through the mediation of chicken wire. He shared how much he appreciated the presence of chickens, especially roosters, in his life, and asked me to let him know if I encountered any other roosters in search of a place to be in the world.

When I think of this story through the hegemonic lens of human mastery, I chastise myself for not being *enough*—careful enough, observant enough, responsible enough—to keep those eggs from ever being incubated. Then, I chastise myself for not being enough once they hatched—resourceful enough, committed enough, smart enough. When I think of this experience from a community-care paradigm, I grieve for the disruption to the cockerels' lives. But alongside that sorrow is a deep appreciation. I feel gratitude that my hens can move more freely through the space they inhabit, and I appreciate how another person helped create the circumstances in which it was possible to take care of—not just take—the lives of three chickens. And for a moment in time in this small part of the web of relations in which we are all entangled, that's enough.

NOTE

1. Of course, fabric does not nourish the land. However, the Project discusses the tons of fabric and clothing that are dumped every year and suggests that repurposing fabric waste into a communal art project is a better fabric-land relation than winding up in a landfill.

Chapter 7

The Final (for Now) Analysis

Which animal does not belong in this list: humans, dogs, cats, goats, squirrels, and slugs. If you selected slugs, that makes sense. They are the only non-mammal in the list. According to both Linnean and cladistic taxonomies,[1] the first five animals are more closely related from an evolutionary perspective. They are morphologically more similar and share a more recent common ancestor. Yet, as I have learned on my small rescue farm, the humble peanut introduces different lines of affinity. The two humans, three dogs, seven goats, and an unknown number of squirrels and slugs, as well as the twenty-two chickens that cohabitate in this patch of Piedmont forest, all enjoy peanuts. The cat does not. The point here is *not* to find which other animals are most like humans and which are most different[2]; the point is to journey with attention to affinities and diversities and what they reveal. Matthew Calarco (2008) points out that a radical openness to these surprising zoographies can help foster appreciation for human *poroi*—human means of passage, ways-of-traveling in this world—in the more-than-human world, including where human paths intersect with other travelers on Earth and where they depart.

What can be taken from the journey of this book is that as humans traverse in more-than-human worlds, including digital spaces, the study and practice of worldbuilding rhetorics—and associated concepts like disclosure, transformation, and infrastructuration—provide some tools for maintaining attention and leaving open possibilities for the *otherwise* in one's feelings, thoughts, beliefs, relationships, and behaviors. Digital worldbuilding rhetorics, like Cary Wolfe's (2010) description of posthumanism, are "not the triumphal surpassing or unmasking of something but an increase in the vigilance, responsibility, and humility that accompany living in a world so newly, and differently, inhabited" (47).

Throughout this book, I have acknowledged that worldbuilding is a rhetorical concept, meaning that suasion participates in the constructions and acts of worlding. I have also stressed that although worldbuilding is a neutral term, the worlds that get built never are. In line with posthumanist and new materialist scholarship, the study of worldbuilding rhetorics is entangled with ethics and politics. Worldbuilding projects occur all the time in both physical and digital spaces, in fictional and nonfictional discourses, and it is important to explore what kinds of communities are being constituted and what modes of care are possible in these projects. Throughout, I have emphasized both *how* the more-than-human world *shows up* in Instagram food photography, Facebook communities, and project websites, and *why* the question of "how does the more-than-human world show up here" is crucial for reminding the *anthropos* that the earth is not a mere backdrop for human affairs but is the *topos*—the commonplace—in and through which all (more-than-)human activity occurs. As Stacey Alaimo puts it,

> the evacuation of agency from nature underwrites the transformation of the world into a passive repository of resources for human use. Alternative conceptions which accentuate the lively, active, emergent, agential aspects of nature foster ethical/epistemological stances that generate concern, care, wonder, respect, caution (or precaution), epistemological humility, kinship, difference, and deviance. (Alaimo 2010, 143)

However, I resist adopting a totalizing "One Earth" rhetoric, in which the *topos* of Earth affords rhetorics of *ecological rationalism* that can be just as totalitarian and damaging as economic rationality (Plumwood 2002). Within the *topos* of the biosphere—this most encompassing of earthly communities—there are many communities and individuals. More-than-human scholarship cannot afford to bracket the *human* in more-than-human and must be attentive to the intersections of sexism and animal oppression (e.g., Adams 2015), the oppressional intersections of race, ethnicity, and animality (e.g., Kim 2015; Ko and Ko 2017), the intersections of LGBTQIA+ (e.g., McHugh 2011), disability (e.g., Taylor 2017), and animal liberation, and the impacts of globalized capitalism on (more-than-)human ways-of-life (e.g., Tsing 2015). As Claire Jean Kim (2015) so carefully illustrates, people may feel compelled—for example—to put a species on the endangered species list to try to save that community of animals from extinction. But what does one do when that endangered species is part of a complex web of relations with Indigenous peoples? Does the government forbid the hunting, prohibiting a way of living-in-the-world that has already been subjected to colonization of land and language? There are no easy answers to this, but Kim describes that any approach must acknowledge both the nonhuman animals' lives and the peoples' lives being affected. One cannot simply have a trump card over the other. Thick, impure engagement must be had among all perspectives, and whatever steps are taken, they must be understood to be revisable.

Yet, the reality and complexity of entanglement does not authorize inaction. Actually, there is no such thing as inaction. In chapter 6, I discussed that—following the work of Karen Barad—*interventions* in the (more-than-)human world are better understood as *intraventions*: humans are not separate beings from the (more-than-)human. Because of human embodiment and embeddedness, all actions occur in a mutual web of relations, and any so-called *inaction* is still a part of this web. Some worlds, as Giraud (2019) notes, are and will be brought into being at the expense of other worlds. This is a simple acknowledgment that lies at the heart of *vulnerable flourishment*—flourishment is partial, provisional, and subject to change. To be committed to (more-than-)human flourishment—the thriving of multiple ways of being-in-the-world—is no small or easy task. It is, through and through, a commitment to *stay with the trouble* (Haraway 2016).

I conclude this book by providing some ways that the rhetorical concepts developed throughout this book—in the context of the community-care paradigm and the wider landscape of rhetorical worldbuilding concepts—can help people stay with the trouble in pedagogy-and-practice (hyphenated in recognition that pedagogy is practice and practice is pedagogical). Pedagogy-and-practice is always participating in worldbuilding rhetorics. This is true whether the processes and methods of instruction pertain to matters of literature, chemistry, or any other discipline. What is enacted and what is taught participate in bringing worlds into being, whether it's a surgical technique that saves lives or an environmental justice concept like *environmental racism* (Bullard 2018) that allows people to understand how toxic waste repositories come to be established in some communities and not others. Here, I elaborate on how disclosure, transformation, and infrastructuration contribute to pedagogical theory and then offer some practical suggestions for how (more-than-)human worldbuilding can be taught.

DISCLOSURE

Disclosure, as discussed in chapter 4, refers to both the purposeful acts of interpersonal communication *and* to rhetorical notions of *the way things show up*. For rhetorical theory, attending to both the interpersonal and ambient aspects of disclosure offers a *multi-optic* framework (Kim 2015) that acknowledges the agency of individuals and the agency that exceeds individuals—even agency that exceeds *bios* altogether. This multi-optic framework helps prevent evacuating agency from the more-than-human world *and* overcorrecting the problem by evacuating agency from humans, which is a hazard of the work that decenters the human. As Nina Lozano (2019) writes, such work "risks rendering unintelligible the ways that individuals and groups working within systems of power communicate and interrelate, within their lived conditions, in collective struggle for social change" (6). Matters of justice—social and ecological, or socioecological—need to be approached with multi-optic lenses not only to ensure mutual (more-than-)human flourishment but also to retain measures of and possibilities for accountability.

As a worldbuilding concept, disclosure can contribute productively to conversations taking place throughout disciplines. For example, this dual understanding of disclosure offers students of design—a discipline concerned with human's adaptations of their surroundings for functional, pleasurable, and ethical purposes (Gallagher, Norris, and Ma 2011)—a way to consider how the intentional practice of place-making and invention interact with the agency of places to afford "visceral, unconscious, socio-historical processes" (Renner 2020, 4). A sustained attention to purpose (e.g., designing for environmental health) *and* to impact (e.g., how designs actually function on/in/with the more-than-human world) is important for the ongoing, inventional work of design. As Joanna Boehnert (2018) describes, design work can help make new ways of living both possible and appealing but requires designers' critical and reflective engagement.

This expanded conceptualization of disclosure also contributes to interpersonal communication theory. The idea that it is impossible not to communicate is a tenet of interpersonal communication, and that includes communicating aspects of one's ecocultural identity. *Ecocultural identity* is a term and a framework for understanding all identities as sociocultural and inseparable from ecological dimensions. "The ecological turn in this conceptualization of identity hinges upon the assumption that *all* identities have earthly constitutions and forces" (Milstein and Castro-Sotomayor 2020, xix). *How ecocultural identities are disclosed* is a new direction for interpersonal pedagogy and research, and the dual understanding of disclosure is useful in exploring this question. One may choose (or not) to purposefully disclose attitudes, beliefs, or values about the more-than-human world, *and* attitudes, beliefs, and values are always *showing up* through actions, words, and relations. The consonance (e.g., eating a vegetarian diet and announcing to friends and family that one has adopted a vegetarian diet) or dissonance (e.g., practicing a vegan diet but choosing not to reveal oneself as vegan) between disclosures (i.e., between verbal statements and what shows up, as discussed in chapter 4) is a fruitful area of relationship-building inquiry for interpersonal communication, regardless whether the content of disclosure is, for example, ecological, political, or biographical.

Further, understanding the relational dynamics that afford and constrain the consonance and dissonance of disclosures makes interpersonal communication an invaluable area of study for environmental studies. Ecocultural identities, as the name implies, are not formed in a vacuum, and interpersonal communication studies can provide important insight into relational and cultural dynamics that encourage or discourage particular identities.

TRANSFORMATION

When phenomena get disclosed *differently,* an ambiguous space is created that invites transformation. In chapter 5, transformation was described as the potential for affective, cognitive, emotional, behavioral, and relational tendencies, as well as spatial arrangements, to change at any moment. The possibility of transformation is what allows conditions or relations of vulnerability to become conditions and relations of flourishment. In short, any practice in which a person imagines, engages, or behaves in ways that "radically restructure our relationships with each other, animals, and the earth outside of domination" (Kim 2015, 21) provides an opening for new ways of being in the world, ways of unmaking the Anthropocene.

Chapter 5 extends the ethnographic methods taken up by scholars engaging in rhetorical field methods by offering digital ethnographic work as another way of exploring routines, socialities, mobilities, practices, values, and relations. Throughout this book and especially in chapter 5, the fluidity of online-offline experience is made apparent through this digital ethnography, and importantly, chapter 5 illustrates how engagement in social media communities can transform attitudes, values, and practices offline, as well. Transformation can and does occur at a distance. This recognition is neither new nor profound; there is substantial discourse on the radicalizing potential of social media platforms. However, chapter 5 reveals the ways in which techniques of using technologies can serve to instrumentalize (and thus marginalize or erase from ethical consideration) *or* particularize (and thus make entities visible and available for ethical consideration) the (more-than-)human world.

Transformation, as discussed in this book, also gestures toward the relevance of interpersonal and organizational communication scholarship for a robust understanding of the (re)shaping of ecological attitudes and values. The transformation of attitudes and actions toward snakes—explored in the chapter—was afforded by the careful crafting of an invitational space. The caring ways in which people's queries were addressed encouraged ongoing involvement with snake identification communities and facilitated more acts of caring. In this way, communication and acts of care can be understood as a practice that locates and embodies care-recipients in a globalized world of dizzying temporalities and scale. Put differently, in the aggregating spaces of Big Data, the careful responses in the snake identification communities allowed "the invisibility" of individuals "to fade away."[3]

Similarly, even seemingly transactional, digital communities like snake identification boards—in which people post a picture to get a fast identification—have the potential to be transformative spaces when they engage in confirming communication practices. In the case of snake identification communities, moderators crafted an invitational space in which commentors did not *endorse* (considered to be the strongest type of confirmation) posts containing images or descriptions of dead snakes, but moderators did craft a space in which people's attitudes and feelings were recognized and acknowledged. Navigating the space of affirmative communication between endorsement and acknowledgment is challenging and each circumstance will present unique needs for this navigation. Thus, chapter 5 highlights how greater collaboration across communication traditions can contribute to more robust theories of ecological and invitational rhetorics.

INFRASTRUCTURATION

Chapter 6 introduced *infrastructuration* as the processes and consequences of ideas, relations, and materialities becoming infrastructure. Infrastructuration is valuable to rhetoric because it animates productive questions like, *What discourses, relations, and ways-of-being are likely to be animated by a particular infrastructure? How might this infrastructure disclose the (more-than-)human world? How and to what extent could this infrastructure enable transformative encounters and conversations? How might this infrastructure organize space and shape mobilities among the (more-than-)human world? In what ways could this infrastructure work with/against a community-care paradigm? What kinds of individuals and communities does this infrastructure invite?*

Infrastructuration makes clear that infrastructure is not only bridges and roads or fiber optic cables and server farms but also organizations and what they organize around, logics and discourses, laws and literacies—and the list goes on. Infrastructuration attends to the inventive agency of more-than-human beings, land, and communication media and technologies to compose pervasive, enabling resources in network form (Bowker et al. 2010). Infrastructuration—as a pedagogical concept—is a way to provoke thinking about what happens when certain entities or phenomena tend to gather with other entities or phenomena. For example, Kristin Arola (2017) argues that Facebook's interface—though brought into being by an often-problematic infrastructure—can be part of decolonial infrastructuration when it is engaged with Indigenous social media techniques and land ethics.

Infrastructuration is a reminder that all disciplines—not just engineering, architecture, and design—are engaged in the process of constructing ways of being in and moving through the (more-than-)human world. There is no absence of an ecocultural identity: an ecocultural identity may be one that dismisses the (more-than-)human world, but that dismissiveness is still an ecocultural identity. This holds not only for individual identities but also for all cultures and forms of human organization. If a syllabus for organizational communication or business ethics, for example, does not include a unit on ecological considerations, that erasure of the more-than-human world contributes to anthropocentric infrastructures that undergird the Anthropocene. Recently, I confess, I engaged in this very practice of teaching organizational communication with no acknowledgment of the wider environment, and I even taught environmental communication *immediately* before teaching a section of organizational communication. I make this confession not out of seeking some sort of absolution but to accentuate how hard it can be to recognize and intervene in infrastructuration.

Finally, although infrastructuration refers to more than roads, bridges, buildings, and power supplies, the study of such material infrastructures is germane across disciplines of study. Bridges and roads are frequently material sites of social and environmental injustice, such as the June 2022 court ruling upholding the transfer of thousands of acres of land sacred for Apaches in Arizona to Rio Tinto Plc to mine copper for the electrification of highways, with the goal to make the road itself the source of energy (Scheyder 2022). Infrastructuration does not provide easy answers to the quandary these copper mines present. The copper mines participate in a network of greener technologies, as the electrification of highways is a means to move away from fossil fuels and reduce the carbon footprint of human mobilities. The copper mines also perpetuate a colonial network of relations and offer what Lauren Berlant (2011) refers to as *cruel optimism*, when something desired—in this case, the continued ease of modern transit—is also an obstacle to flourishment. Consider that the mining of copper that may abate some of the damage done by fossil fuels also hampers restorative relations with Indigenous communities and perpetuates a perspective of land as part of commodity and not community. *Infrastructuration* is a concept that invites multidisciplinary answers to the questions of *what kind of world does a particular infrastructure bring into being* and *how does this infrastructure afford and constrain living well with (more-than-)human others?*

TOWARD ECOLOGICAL READINESS

At best, the concepts in this book can help facilitate a sense of *ecological readiness* in students and others. Like Danielle Allen's (2016) notion of *participatory readiness* for civic affairs, *ecological readiness* is a sense of being able to engage in acts of (more-than-)human worldbuilding at levels from the home to the global village. An ecologically ready participant is someone ready to participate in caring for the more-than-human world, one who is ready to recognize oneself as a member of a more-than-human community. An ecologically ready participant is one who understands the complexity and entanglement of these acts and is prepared to participate in cocreating worlds of multispecies flourishment.

Ecological readiness can only be made more or less probable—not guaranteed—by education; it is not a learning objective but a state of thinking-feeling-acting. Educators and practitioners can invite a sense of readiness through learning objectives, such as teaching students learn to identify and analyze how discourse constructs relationships between the more-than-human world and humans and helping students gain skills in oral, written, and visual communication in multimodalities such that they have a sense of efficacy in their capacity to act. Such learning objectives afford the *dunamis*—the potential—for ecological readiness to flourish, but they are not a sufficient *cause* of readiness, only an inducement.

Ecological readiness exists on a spectrum and in ecologic relations. In other words, ecological readiness is not a binary state of being ready/not ready but exists in degrees, and the extent of readiness is impacted by a complex ecology of relations, such as health, current living conditions, awareness of issues, opportunities for participation, the health of the political climate, the existence or absence of infrastructure that supports participation. There are no guarantees that an ecologically ready participant will be met with a receptive public sphere. However, there are some insights from rhetoric and ethics that can help prepare individuals to see and act on opportunities when they arise—including the capacity to recognize opportunities that exist beyond the public sphere.

The notion of worldbuilding can help people attend to the fluidity, relationality, and more-than-human aspects of rhetoric and of ethics of care. Understood from the worldbuilding perspective of the community-care paradigm, *rhetoric* is less an action (e.g., the delivering of a speech, the asking of a performative question) than a (more-than-)human force that invites different ways of relating and moving in the world, and an *ethics of care* includes processes, acts, and feelings that are always influenced by (more-than-)human rhetoric. The active nature of worldbuild*ing* can invite paralysis of action, as one can become cognizant of the continual unfolding and unknowability of the consequences of one's actions. But the verb can also invite creative action, free from the constraint of believing that the "state of things" is a settled and done matter. Philosophers of ethics disagree on many things, but a widely shared view is that any statement of "ought"—how one "ought" to respond to or act in a situation—implies *can*.[4] By inviting people to see things as processes than as stabilized nouns, the notion of *worldbuilding* helps open up the possibilities of what *can* be done, and the recognition of the *vulnerable flourishment* of these worlds-in-the-making necessitates these possibilities be approached with humility.

The heart of this book is an exploration of how rhetoric functions to disclose, transform, and build infrastructure in digital communities. This book is just the tip of the metaphorical iceberg. As humans increasingly dwell in digital spaces, it is important that substantial attention be devoted to how to embody and emplace those spaces. Ecologically ready participants must first recognize themselves as participants in an ecology. Such a recognition requires not approaching a cell phone as a *black mirror*[5] whose test humans and other animals either pass or fail, but as a technology that participates in disclosing, transforming, and enabling different worlds. The Anthropocene announces the precarity of the world in which digital technologies are embedded, but that world is still being built and can be *otherwise*.

While I feel certain that none of what is written in this book is any great revelation, my recognition of and willingness to jump into adjacent possibilities may be productive. The community-care paradigm is developed through deep engagement with ethics of care and ecological theory, but it is not the birth of something new; it is a relative of Indigenous ontology, ethics, and epistemology. The basic unit I identify of individuals-in-communities is also a key feature of many Indigenous cosmovisions and new materialist rhetorics. A recurrent concept in this book, *vulnerable flourishment*, is a recognition of the entangled states of precarity and wellbeing, which is news to neither interpersonal communication scholars, social work practitioners, nor ecological theorists. The worldbuilding capacity of rhetoric was begrudgingly disclosed by Plato (2001) thousands of years ago and has been explicitly labelled and theorized in the twenty-first century by scholars like Thomas Rickert (2013) and Jennifer Clary-Lemon (2019).

Disclosure is both a key concept of every entry-level interpersonal communication classes, and also a recent addition to rhetorical theory through the Heideggerian interpretation offered by Rickert (2013). *Transformation* shows up everywhere from techniques of motivational interviewing that occasion change in a person's life (Miller and Rollnick 2012) to Deleuze and Guattari's (1987) explication of the ever-unfolding process of *becoming*. *Infrastructuration* is built on STS and informatic studies of *infrastructuring* (Karasti, Pipek, and Bowker 2018) and is evocative of Marxist theory of historical materialism (though without economics and class distinction as the archetypal infrastructure) and Foucault's archaeology of knowledge (1982) that queries the material conditions of discourse.

Thus, nothing here is a particular revelation, nor should anything written here be interpreted as offering a panacea. Although I am loathe to constrain interpretations, the affirmative approach taken in this book *should not* be read as a utopic vision. The disclosures offered by vernacular plant-based visual rhetorics through the digital food photography practices on Instagram are not going to usher in some age of mass conversion to plant-based diets and provoke widespread consideration of what it means to *eat well* (Derrida 1991). The transformation of attitudes toward snakes afforded by snake identification communities on Facebook may be insufficient to combat the threats to snakes' lives posed by habitat destruction and climate change. The infrastructuration of projects like Reclaim the Void that create decolonial networks of relations for living in and with the land are not stopping the robust mining infrastructure in Australia, the United States, or anywhere that "resources" can be found and extracted.

So, what does this book do? It affirms that the Anthropocene is not determined by being an *anthropos* but by certain ways of being an *anthropos* that have been made and can be unmade. It recognizes that although the evidence of the harmful impacts of social media platforms is bountiful—from the data capitalism it brings into being to the often devastating personal and social impacts of everything from privacy violations to radicalization to dangerous viral "challenges" to a *fear of missing out* now simply known as FOMO—it is not determined to be so. The same extractive logics and ethics that mark the Anthropocene's material alteration of the more-than-human physical world are also at play in more-than-human digital spaces, and this book functions as a reminder that there are ways of being *otherwise* in both. There are ways of being human that are affirmative of the rich web of relations in which the species exists. The Anthropocene amplifies (more-than-)human vulnerability, but it also heightens the exigence for rhetorics that contribute to (more-than-)human flourishment. This book is an attempt to document some of the everyday practices, projects, and techniques that stay with the trouble of living well with others in physical and digital spaces.

STORY TIME: FALL BLOOMS

Fall 2022: For the first time in the six years since we moved into our wooded six-acre farm, there are chrysanthemums blooming in the flower bed that lines the front of our house. They are a shade of orange evocative of the autumn season, and I have an affection for flowers that bloom while the leaves of deciduous trees change into their fall colors. The late bloomers remind me that flourishment is a kairotic affair, occurring in its own time.

Beautiful, yes, and there's more to these petals. There is this plant's own origin story, which is unknown and likely unknowable to me. There is also the story of how the plant came to reside in the land I do my best to steward, and on that front, I know a bit more.

On the last Thursday of September, I apologized to my communication media class. For most of the month of September, I had been teaching classes and attending health care appointments with my partner. He had had an accident that resulted in a traumatic brain injury (TBI), and the consequences of that trauma unfolded over months with multiple trips to the emergency room, a multi-day hospitalization, and numerous outpatient appointments for lab work and imaging (he tells me he is certain he holds the record for most CT scans in a several-week period). During this time, a tree also fell and took out part of the fence that contains our goats, one of our dogs had a major surgery that required extensive postoperative care, and I became acquainted with how much my partner and I divide the routine and daily care of the house when I needed to do those activities solo.

Although I was actively writing about vulnerable flourishment, I found precious little flourishment in my own life. The actions in which I had been intentionally engaging for my wellbeing—running, playing guitar, spending time with the companion critters outside of attending to basic life requirements—had ceased, and I was not showing up in my classroom with the presence that I value bringing as an instructor.

On that final Thursday of September, I decided to lean into my own vulnerability. In what could not have taken more than two minutes, I began class by apologizing for not being as present as I value being in the classroom and thanking them for their adaptability as we made changes and for their grace when I stumbled over a joke or remembering our schedule. The following Tuesday, I was greeted by a pot of chrysanthemums and a card that read, "Nice things just seem to bloom from you." I had not told my students about this book, about the community-care paradigm, or about the notion of *vulnerable flourishment*, but there it was. The flowers happened to share our classroom space on a day that we were doing an actor-network theory activity in class. I showed them a network visual of my life that I made as an example. As the flowers sat in front of the projected network visualization, I was struck by my own naivete on display in that visualization. The only link I had drawn between my house and my university's campus was my Prius—the transportation medium that carries me between the two. But of course those nodes were always more densely connected than that! My partner's TBI was never confined to the home, and the content of my course and my students' wellbeing did not leave my mind as I got into my car. My life is not an episode of *Severance*, the Apple TV series in which a corporation has its employees' brains modified such that they do not have access to memories of work when at home or memories of home when at work.

As I carried the flowers home in my Prius, I thought about how the flows of community-care were showing up in my life. For most of 2022, I have felt vulnerable in all directions. And I am. My home is vulnerable, my partner is vulnerable, my companion critters are vulnerable, I am vulnerable. And that vulnerability also is the impetus for mundane but profound, ongoing acts of care. I care for my companion chickens by providing food and shelter, and they also provide care for me through the daily sustenance of their eggs and through the joy of sharing space with them. With that nourishment, I go to campus to teach—a caring of its own—and then I return home nourished by the ideas, the energy, the creativity, and the kindness of my students. Vulnerable flourishment extends in all directions.

Spinoza (1996) said that nonhuman animals do not agree with us in nature, but he also said that we do not even know what a body is capable of, of what affections humans are capable. I think we should consider that first statement in light of the second: We do not even know what humans can do, of what affections we are capable, and in what (more-than-)human affective communities we can be present and how. The kinds of care I provide and are provided to me by chickens, dogs, and other humans are distinct, but I could not begin to rank them. I am grateful for the possibilities of living well with others that each relationship makes available. Each relationship helps me flourish, to live more fully as a being on Earth, etc.[6]

NOTES

1. The Linnean and cladistic systems are two different taxonomies for living beings based on different methodologies and paradigms. The Linnean system is a hierarchical system of classifying animals according to shared characteristics. The cladistic system focuses on the shared ancestry of evolution. Broadly speaking, the Linnean system is a taxonomy that tells us how living beings are *like* other living beings, while the cladistic system is a phylogeny that tells us how living beings are related to other beings. For example, in the Linnean system, birds and reptiles are separate classes based on their distinguishing features (e.g., body covering, thermoregulation), whereas the cladistic system groups them in the same clade—Sauropsids—that recognizes their shared ancestral lineage.

2. This is what Matthew Calarco (2020) calls going *beyond* the anthropological difference. The point is not to find what animals are "most like" us and therefore have our systems of ethics *extended* to them, nor is it to assert other animals' value through their radical otherness. Rather, being beyond the anthropological difference is being open to points of convergence and distinction among all creatures.

3. When I acknowledge that careful responses work against the sweeping generalities of Big Data, I am not suggesting that these responses exist *outside* of Big Data's aggregation. As discussed in chapter 5, algorithms are significant agents in the snake

identification networks and have the capacity to amplify or limit the potential of such communities. Careful responses do not *undo* the datafication (and commodification of that data) of interactions, but they do briefly announce the *individual* as a node or agent in this massive network.

4. Given my blending of ethical approaches, the principle that *ought implies can* warrants some unpacking here. *Ought implies can*, attributed to Kant, is a principle of ethics as an area of philosophic study—that is, it is a principle of the branch of knowledge that deals with moral principles. However, there exists debate about the nature of morality and the nature of ethics. Some use the words interchangeably. Others use *morality* to refer to guiding principles from a higher authority (and are *right* or *wrong*) and *ethics* as immanent, social standards that are *good* or *bad*.

5. *Black Mirror* is a British anthology series depicting technologically induced dystopian near-futures. The title refers to the way the black screens through which one interfaces with many digital spaces also reflect the user.

6. This book has attempted to be an inquiry in (more-than-)human modes and relations of flourishment that include digital rhetorics. As Bruno Latour (2007) writes, inquiries always come in in the middle and produce a lot of new descriptions—thus, all inquiries should end with *et cetera*, acknowledging their ongoing unfolding, acknowledging the adjacent possibilities.

References

Abbey, Edward. 2006. *The Monkey Wrench Gang.* New York: Harper Perennial Modern Classics.

Abram, David. 1996. *The Spell of the Sensuous: Perception and Language in a More-Than-Human World.* New York: Vintage Books.

Adams, Carol. 2015. *The Sexual Politics of Meat: A Feminist-Vegetarian Critical Theory.* New York: Bloomsbury.

Adewale, Omowale, ed. 2021. *Brotha Vegan: Black Men Speak on Food, Identity, Health, and Society.* Brooklyn, NY: Lantern Publishing & Media.

Agamben, Giorgio. 2004. *The Open*. Stanford, CA: Stanford University Press.

Aguilera-Carnerero, Carmen, and Margarita Carretero-González. 2021. "The Vegan Myth: The Rhetoric of Online Anti-Veganism." In *The Routledge Handbook of Vegan Studies*, edited by Laura Wright, 354–66. New York: Routledge.

Ahmed, Sara. 2004. *The Cultural Politics of Emotion*. New York: Routledge.

Alaimo, Stacy. 2010. *Bodily Natures: Science, Environment, and the Material Self.* Bloomington: Indiana University Press.

Allen, Danielle. 2016. "What Is Education For?" *Boston Review,* April 26, 2016. https://bostonreview.net/forum/etflix-allen-what-is-education-for

Aridi, Rasha. 2021. "Lobsters, Crabs, and Octopuses Will Now Receive Welfare Protection as 'Sentient Beings' in the U.K." *Smithsonian Magazine,* November 23, 2021. https://www.smithsonianmag.com/smart-news/lobsters-crabs-and-octopuses-will-be-recognized-as-sentient-beings-in-the-uk-warranting-welfare-protections-180979113/

Aristotle. 1991. *On Rhetoric: A Theory of Civic Discourse*. Translated by George A. Kennedy. New York: Oxford University Press.

———. 2011. *Nicomachean Ethics*. Translated by Robert Bartlett and Susan Collins. Chicago: The University of Chicago Press.

Arola, Kristin L. 2017. "Indigenous Interfaces." In *Social Writing/Social Media: Publics, Presentations, and Pedagogies*, edited by Douglas M. Walls and Stephanie Vie, 211–26. Fort Collins, CO: WAC Clearinghouse.

———. 2018. "A Land-Based Digital Design Rhetoric." In *The Routledge Handbook of Digital Writing and Rhetoric*, edited by Jonathan Alexander and Jacqueline Rhodes, 199–213. New York: Routledge.

Australians Together. 2022. "Colonisation: Dispossession, Disease, and Direct Conflict." *Australians Together.* https://www.australianstogether.org.au/discover/etflixca-history/etflixcan/

Balthrop, V. William, Carole Blair, and Neil Michel. 2010. "The Presence of the Present: Hijacking 'The Good War'?" *Western Journal of Communication* 74, no. 2: 170–207.

Barad, Karen. 2007. *Meeting the Universe Halfway: Quantum Physics and the Entanglement of Matter and Meaning.* Durham, NC: Duke University Press.

Barber, Aja. 2021. *Consumed: The Need for Collective Change: Colonialism, Climate Change, and Consumerism*. New York: Balance.

Barnes, Marian, Tula Brannwlly, Lizzie Ward, and Nicki Ward. 2015. "Introduction." In *Ethics of Care: Critical Advances in International Perspective,* edited by Marian Barnes, Tula Brannwlly, Lizzie Ward, and Nicki Ward, 3–20. Chicago: Policy Press.

Barnett, Joshua Trey. 2019. "Naming, Mourning, and the Work of Earthly Coexistence." *Environmental Communication* 13, no. 3: 287–99.

Bekoff, Marc. 2007. *The Emotional Lives of Animals.* Novato, CA: New World Library.

Bennett, Carys E., Richard Thomas, Mark Williams, Jan Zalasiewicz, Matt Edgeworth, Holly Miller, Ben Coles, Alison Foster, Emily J. Burton, and Upenyu Marume. 2018. "The Broiler Chicken as a Signal of a Human Reconfigured Biosphere." *Royal Society Open Science* 5, no. 12: 180325.

Bentham, Jeremy. 2007. "Principles of Morals and Legislation." In *The Animals Reader: The Essential Classic and Contemporary Writings,* edited by Linda Kalof and Amy Fitzgerald, 8–9. New York: Berg.

Berchin, Issa Ibrahim, Isabela Blasi Valduga, Jéssica Garcia, and José Baltazar Salgueirinho Osório de Andrade. 2017. "Climate Change and Forced Migrations: An Effort Towards Recognizing Climate Refugees." *Geoforum* 84: 147–50.

Bergman, Jordanna N., Rachel T. Buxton, Hsien-Yung Lin, Magdalena Lenda, Kayla Attinello, Adrianne C. Hajdasz, Stephanie A. Rivest, Thuong Tran Nguyen, Steven J. Cooke, and Joseph R. Bennett. 2022. "Evaluating the Benefits and Risks of Social Media for Wildlife Conservation." *FACETS* 7, no. 1: 360–97.

Bergström, Johanna. 2018. "Learning from Mayan Feminists' Interpretations of Buen Vivir." In *Unsettling Eurocentrism in the Westernized University*, edited by Julie Cupples and Ramon Grosfoguel, 156–70. New York: Routledge.

Berkes, Fikret. 1993. "Traditional Ecological Knowledge in Perspective." In *Traditional Ecological Knowledge: Concepts and Cases*, edited by Julian T. Inglis, 1–10. Ottawa: International Development Research Centre.

Berland, Jody. 2009. *North of Empire: Essays on the Cultural Technologies of Space*. Durham: Duke University Press.

———. 2019. *Virtual Menageries: Animals as Mediators in Network Cultures*. Cambridge, MA: MIT Press.

Berlant, Lauren. 2011. *Cruel Optimism.* Durham, NC: Duke University Press.

Beyond Meat. 2022. *Beyond Meat*. Accessed July 1, 2022. https://www.beyondmeat.com/en-US/

Blair, Carole. 1999. "Contemporary U.S. Memorial Sites as Exemplays of Rhetoric's Materiality." In *Rhetorical Bodies*, edited by Jack Selzer and Sharon Crowley, 16–57. Madison: The University of Wisconsin Press.

Blair, Carole, and Neil Michel. 2007. "The AIDS Memorial Quilt and the Contemporary Culture of Public Commemoration." *Rhetoric and Public Affairs* 10, no. 4: 595–626.

Boehnert, Joanna. 2018. *Design, Ecology, Politics: Towards the Ecocene.* New York: Bloomsbury.

Bogost, Ian. 2010. *Persuasive Games: The Expressive Power of Videogames*. Cambridge, MA: MIT Press.

Boisseron, Bénédicte. 2018. *Afro-Dog: Blackness and the Animal Question*. New York: Columbia University Press.

Bolderdijk, Jan Willem, and Gert Cornelissen. 2022. "'How Do You Know Someone's Vegan?' They Won't Always Tell You. An Empirical Test of the Do-Gooder's Dilemma." *Appetite* 168: 1–7.

Bollmer, Grant. 2018. *Theorizing Digital Cultures*. Thousand Oaks, CA: Sage.

———. 2019. *Materialist Media Theory: An Introduction*. New York: Bloomsbury Publishing.

Bone, Jennifer Emerling, Cindy L. Griffin, and T. M. Linda Scholz. 2008. "Beyond Traditional Conceptualizations of Rhetoric: Invitational Rhetoric and a Move Toward Civility." *Western Journal of Communication* 72, no. 4: 434–62.

Bowker, Geoffrey C. 1994. *Science on the Run: Information Management and Industrial Geophysics at Schlumberger, 1920–1940.* Cambridge, MA: Harvard University Press.

Bowker, Geoffrey C., Karen Baker, Florence Millerand, and David Ribes. 2010. "Toward Information Infrastructure Studies: Ways of Knowing in a Networked Environment." In *International Handbook of Internet Research*, edited by Jeremy Hunsinger, Lisbeth Klastrup, and Matthew Allen, 97–117. Dordrecht: Springer Science+Business Media.

Bresnahan, M., J. Zhuang, and X. Zhu. 2016. "Why Is the Vegan Line in the Dining Hall Always the Shortest? Understanding Vegan Stigma." *Stigma and Health* 1, no. 1: 3–15.

Britannica, The Editors of Encyclopaedia. 2013. "Cladoselache." *Encyclopedia Britannica*. https://www.britannica.com/animal/Cladoselache

Broad, Garrett M. 2016. "Animal Production, Ag-Gag Laws, and the Social Production of Ignorance: Exploring the Role of Storytelling." *Environmental Communication* 10, no. 1: 43–61.

Brookshire, Bethany. 2022. *Pests: How Humans Create Animal Villains.* New York: Ecco.

Brosnan, Sarah, and Frans de Waal. 2003. "Monkeys Reject Unequal Pay." *Nature* 425, no. 6955: 297–99.

Brown, Brené. 2012. *Daring Greatly: How the Courage to Be Vulnerable Transforms the Way We Live, Love, Parent, and Lead.* New York: Penguin Books.

Bugalho, Paulo, and Teresa Paiva. 2011. "Dream Features in the Early Stages of Parkinson's Disease." *Journal of Neural Transmission* 118, no. 11: 1613–19.

Bullard, Robert. 2018. *Dumping in Dixie: Race, Class, and Environmental Quality.* New York: Routledge.

Burghardt, Gordon. 2016. "Mediating Claims Through Critical Anthropomorphism." *Animal Sentience* 1, no. 3: 17.

Burke, Kenneth. 2012. "Literature as Equipment for Living." In *The Routledge Reader in Rhetorical Criticism*, edited by Brian L. Ott and Greg Dickinson, 259–63. New York: Routledge.

Burke, Kenneth. 1951. "Rhetoric—Old and New." *Journal of General Education* 5, no. 3: 202–9.

———. 1962. *A Grammar of Motives and a Rhetoric of Motives.* Cleveland, OH: Meridian Books, 1962.

———. 1966. *Language as Symbolic Action: Essays on Life, Literature, and Method.* Berkeley: University of California Press.

———. 1984. *Permanence and Change: An Anatomy of Purpose*, 3rd ed. Berkeley: University of California Press.

Buttny, Richard, and Etsuko Kinefuchi. 2020. "Vegans' Problem Stories: Negotiating Vegan Identity in Dealing with Omnivores." *Discourse & Society* 31, no. 6: 565–83.

Calarco, Matthew. 2008. *Zoographies: The Question of the Animal from Heidegger to Derrida*. New York: Columbia University Press.

———. 2014. *Thinking Through Animals: Identity, Difference, Indistinction.* Stanford, CA: Stanford University Press.

———. 2020. *Beyond the Anthropological Difference.* New York: Cambridge University Press.

———. 2021. *The Boundaries of Human Nature: The Philosophical Animal from Plato to Haraway*. New York: Columbia University Press.

Callicott, J. Baird. 2017. "The Conceptual Foundations of the Land Ethic." In *Environmental Ethics: Readings in Theory and Application* 7th ed., edited by Louis P. Pojman, Paul Pojman, and Katie McShane, 248–60. Boston: Cengage.

Campbell, Stephen. 2016. "The Concept of Well-Being." In *The Routledge Handbook of Philosophy of Well-Being*, edited by Guy Fletcher, 402–14. New York: Routledge.

Camposano, C. C. 2018. "When Objects Speak Louder Than Words: Food, Intimacy and Power in the Contemporary Transnational Filipino Household." In *Food Parcels in International Migration,* 73–93. Cham: Palgrave Macmillan.

Card, Claudia. 1990. "Caring and Evil." *Hypatia* 5, no. 1: 101–8.

Carpenter, Rick. 2013. "Place-Identity and the Socio-spatial Environment." In *Environmental Rhetoric and Ecologies of Place,* edited by Peter N. Goggin, 199–216. New York: Routledge.

Carrington, Damian. 2016. "The Anthropocene Epoch: Scientists Declare Dawn of Human-Influenced Age." *The Guardian*, August 29, 2016. https://www.theguardian.com/environment/2016/aug/29/declare-anthropocene-epoch-experts-urge-geological-congress-human-impact-earth

Carver, Raymond. 1989. *What We Talk about When We Talk about Love.* New York: Vintage Books.

Catching Killers, S1.1, "Body Count: The Green River Killer." Directed by Suemay Oram, aired November 4, 2021, Netflix.

Challenger, Melanie. 2018. "Fear Itself." *Orion.* https://orionmagazine.org/article/fear-itself/

Chalmers, David. 2022. *Reality+: Virtual Worlds and the Problems of Philosophy.* New York: W. W. Norton & Company.

Charland, Maurice. 1987. "Constitutive Rhetoric: The Case of the Peuple Quebecois." *Quarterly Journal of Speech* 73, no. 2: 133–50.

Cheney, George. 1983. "The Rhetoric of Identification and the Study of Organizational Communication." *Quarterly Journal of Speech* 69, no. 2: 143–58.

Chilisa, Bagele. 2019. *Indigenous Research Methodologies*, 2nd ed. Los Angeles: Sage.

Chouliaraki, Lilie, and Angelos Kissas. "The Communication of Horrorism: A Typology of ISIS Online Death Videos." 2018. *Critical Studies in Media Communication* 35, no. 1: 24–39.

Clary-Lemon, Jennifer. 2019. *Planting the Anthropocene: Rhetorics of Natureculture.* Louisville: University of Colorado Press.

———. 2022. "From Precarity to Invasion: Nonhuman Rhetorics in Times of Global Climate Change." Presentation at the 20th Biennial Conference of the Rhetoric Society of America, Baltimore, MD, May 26–29, 2022.

Cole, Adam, producer. 2019. "Memory." *The Mind, Explained.* Netflix. https://www.netflix.com/watch/81062188

Collins, Patricia Hill. 2000. *Black Feminist Thought.* New York: Routledge.

Condit, Celesta Michelle. 1997. "In Praise of Eloquent Diversity: Gender and Rhetoric as Public Persuasion." *Women's Studies in Communication* 20, no. 2: 91–116.

Cooke, Lucy. 2022. *Bitch: On the Female of the Species.* New York: Basic Books.

Cox, J. Robert. 1982. "The Die Is Cast: Topical and Ontological Dimensions of the Locus of the Irreparable." *Quarterly Journal of Speech* 68, no. 3: 227–39.

Craig, Elizabeth A., Leanne Nieforth, and Cynthia Rosenfeld. 2020. "Communicating Resilience among Adolescents with Adverse Childhood Experiences (ACEs) through Equine Assisted Psychotherapy (EAP)." *Western Journal of Communication* 84, no. 4: 400–418.

Crary, Jonathan. 2022. *Scorched Earth: Beyond the Digital Age to a Post-Capitalist World.* Brooklyn, NY: Verso.

Curry, Patrick. 2011. *Ecological Ethics.* Malden, MA: Polity.

Dadas, Caroline. 2017. "Hashtag Activism: The Promise and Risk of 'Attention.'" In *Social Writing/Social Media: Publics, Presentations, and Pedagogies*, edited by Douglas M. Walls and Stephanie Vie, 17–36. Fort Collins, CO: WAC Clearinghouse.

Daltry, Jenny C., Quentin Bloxam, Gillian Cooper, Mark L. Day, John Hartley, McRonnie Henry, Kevel Lindsay, and Brian E. Smith. 2001. "Five Years of Conserving the 'World's Rarest Snake,' the Antiguan Racer Alsophis Antiguae." *Oryx* 35, no. 2: 119–27.

Darmiento, Laurence. 2020. "Ethan Brown Went Vegan but Missed Fast Food. So He Started a Revolution." *Los Angeles Times,* January 10, 2020. https://www.latimes.com/business/story/2020-01-08/beyond-meat-founder-ethan-brown

Davis, Diane. 2011. "Creaturely Rhetorics." *Philosophy & Rhetoric* 44, no. 1: 88–94.

———. 2014. "Autozoography: Notes Toward a Rhetoricity of the Living." *Philosophy & Rhetoric* 47, no. 4: 533–553.

de Kosnik, Abigail, and Keith P. Feldman. 2019. *#identity: Hashtagging, Race, Gender, Sexuality, and the Nation*. Ann Arbor: University of Michigan Press.

de la Bellacasa, Maria Puig. 2017. *Matters of Care: Speculative Ethics in More Than Human Worlds.* Minneapolis: University of Minnesota Press.

de Waal, Frans. 1999. "Anthropomorphism and Anthropodenial: Consistency in Our Thinking about Humans and Other Animals." *Philosophical Topics* 27, no. 1: 255–80.

Deckha, Maneesha. 2018. "Postcolonial." In *Critical Terms for Animal Studies,* edited by Lori Gruen, 280–93. Chicago: University of Chicago Press.

Deecke, Volker B., John K. B. Ford, and Paul Spong. 1999. "Quantifying Complex Patterns of Bioacoustic Variation: Use of a Neural Network to Compare Killer Whale (*Orcinus orca*) Dialects." *The Journal of the Acoustical Society of America* 105, no. 4: 2499–2507.

Deleuze, Gilles. 1988. *Spinoza: Practical Philosophy.* San Francisco: City Light Books.

Deleuze, Gilles, and Felix Guattari. 1987. *A Thousand Plateaus*. Translated by Brian Massumi. Minneapolis: University of Minnesota Press.

———. 2009. *Anti-Oedipus: Capitalism and Schizophrenia*. Translated by Robert Hurley, Mark Seem, and Helen Lane. New York: Penguin Books.

DeLuca, Kevin. 1999. *Image Politics: The New Rhetoric of Environmental Activism*. New York: Routledge.

Derrida, Jacques. 1991. "'Eating Well,' or the Calculation of the Subject: An Interview with Jacques Derrida." In *Who Comes After the Subject?,* edited by Eduardo Cadava, Peter Connor, and Jean-Luc Nancy, 96–119. New York: Routledge.

Dery, Mark. 1990. "The Merry Pranksters and the Art of the Hoax." *New York Times,* December 23, 1990. https://www.nytimes.com/1990/12/23/arts/the-merry-pranksters-and-the-art-of-the-hoax.html

———. 1993. *Culture Jamming: Hacking, Slashing and Sniping in the Empire of Signs*. Open Magazine Pamphlet Series. https://www.markdery.com/books/culture-jamming-hacking-slashing-and-sniping-in-the-empire-of-signs-2/

Diamond, Cora. 2003. "The Difficulty of Reality and the Difficulty of Philosophy." *Partial Answers: Journal of Literature and the History of Ideas* 1, no. 2: 108–28.

Diamond, Jared. 2005. *Collapse: How Societies Choose to Fail or Succeed.* New York: Viking.

Dickinson, Emily. 2011. "Displaced in Nature: The Cultural Production of (Non-)Place in Place-Based Forest Conservation Pedagogy. *Environmental Communication* 5, no. 3: 300–319.

Dickinson, Emily. "A Narrow Fellow in the Grass," *Poetry Foundation*, accessed June 29, 2022. https://www.poetryfoundation.org/poems/49909/a-narrow-fellow-in-the-grass-1096

Dickinson, Greg. 2002. Joe's Rhetoric: Finding Authenticity at Starbucks. *Rhetoric Society Quarterly* 32, no. 4: 5–27.
Dickinson, Greg, Brian L. Ott, and Aoki Eric. 2005. "Memory and Myth at the Buffalo Bill Museum." *Western Journal of Communication* 69, no. 2: 85–108.
Dickinson, Greg, and Casey Malone Maugh. 2004. "Placing Visual Rhetoric: Finding Material Comfort in Wild Oats Market." In *Defining Visual Rhetorics,* edited by Charles A. Hill and Marguerite Helmers, 259–76. Mahwah, NJ: Lawrence Erlbaum Associates, Inc.
Donald, Dwayne. 2010. "On What Terms Can We Speak?" Lecture at the University of Lethbridge. www.vimeo.com/15264558
Donovan, Josephine, and Carol Adams. 2007. "Introduction." In *The Feminist Care Tradition in Animal Ethics*, edited by Josephine Donovan and Carol Adams, 1–20. New York: Columbia University Press.
Duncan, Dennis. 2022. *Index, A History of the.* New York: W. W. Norton & Company.
Durso, Andréw Michaël, Isabelle Bolon, A. R. Kleinhesselink, M. R. Mondardini, J. L. Fernandez-Marquez, F. Gutsche-Jones, C. Gwilliams, M. Tanner, C. E. Smith, W. Wüster, F. Grey, and R. Ruiz de Castañeda. 2021. "Crowdsourcing Snake Identification with Online Communities of Professional Herpetologists and Avocational Snake Enthusiasts." *Royal Society Open Science* 8, no. 1: 1–21.
Ehninger, Douglas. 1968. "On Systems of Rhetoric." *Philosophy & Rhetoric* 1, no. 3: 131–44.
Elder, John. 2014. "Foreword." In *Matters & Desire: An Erotic Ecology*, edited by Andreas Weber, ix–xii. White River Junction, VT: Chelsea Green Publishing.
Endres, Danielle, Aaron Hess, Samantha Senda-Cook, and Michael K. Middleton. 2016. "*In Situ* Rhetoric: Intersections Between Qualitative Inquiry, Fieldwork, and Rhetoric." *Cultural Studies? Critical Methodologies* 16, no. 6: 511–24.
Environmental Protection Agency. 2006. "Permethrin Facts." *Pesticides*. https://www3.epa.gov/pesticides/chem_search/reg_actions/reregistration/fs_PC-109701_1-Jun-06.pdf
Erben, Şeyma Esin, and Jale Balaban-Sali. 2016. "Social Media Sse of Vegan Activists in Turkey." *Global Media Journal* 6, no. 12: 75–88.
Estok, Simon. 2021. "Merchandizing Veganism." In *The Routledge Handbook of Vegan Studies*, edited by Laura Wright, 333–42. New York: Routledge.
Feldman, Abigail. 2021. "What Is an Aesthetic of Access?" In *Immersive Scholar: A Guidebook for Documentation and Publishing Experiential Scholarship Works*, edited by Abigail E. Mann and Micah Vandegrift, 7–8. Raleigh: North Carolina State University Libraries.
50 Words Project. *50 Words Project*, accessed July 2, 2022. https://50words.online/
Figueres, Christiana, and Tom Rivett-Carnac. 2020. *The Future We Choose: Surviving the Climate Crisis*. New York: Vintage.
Find & Connect. "Colony of Western Australia (1829–1901)." *Find & Connect*, accessed September 15, 2023. https://www.findandconnect.gov.au/guide/wa/WE01504
Finnegan, Cara A. 2005. "Recognizing Lincoln: Image Vernaculars in Nineteenth-Century Visual Culture." *Rhetoric & Public Affairs* 8, no. 1: 31–57.

Fisher, Carl Erik. 2022. "It's Misleading to Call Addiction a Disease." *New York Times,* January 15, 2022. https://www.nytimes.com/2022/01/15/opinion/addiction-disease.html

Flavelle, Christopher, Denise Lu, Veronica Penney, Nadja Popovich, and John Schwartz. 2020. "New Data Reveals Hidden Flood Risk Across America." *New York Times*, June 29, 2020. https://www.nytimes.com/interactive/2020/06/29/climate/hidden-flood-risk-maps.html

Flavelle, Christopher, and Nadja Popovich. 2022. "Here Are the Wildfire Risks to Homes Across the Lower 48 States." *New York Times*, May 16, 2022. https://www.nytimes.com/ interactive/2022/05/16/climate/wildfire-risk-map-properties.html

Flitter, Emily. 2022. "It's Hard to Tell When the Crypto Bubble Will Burst, or If There Is One." *New York Times*, January 27, 2022. https://www.nytimes.com/2022/01/27/business/crypto-price-bubble.html

Foer, Jonathan Safran. 2009. *Eating Animals.* New York: Little, Brown and Company.

———. 2019. *We Are the Weather: Saving the Planet Begins at Breakfast*. New York: Farrar, Straus and Giroux.

Fonseca, Felipe Schmidt. 2017. "Knowledge, Skill, and Labor." *Medium,* January 13, 2017. https://felipefonseca.medium.com/knowledge-skill-and-labor-1c8f5d89a1b4

Forbes, Jack D. 2001. "Nature and Culture: Problematic Concepts for Native Americans." In *Indigenous Traditions and Ecology: The Interbeing of Cosmology and Community*, edited by John Grim, 103–24. Cambridge, MA: Harvard Divinity School.

Foss, Sonja K., and Cindy L. Griffin. 1995. "Beyond Persuasion: A Proposal for an Invitational Rhetoric." *Communications Monographs* 62, no. 1: 2–18.

———. 2020. "Proposal for a Feminist Rhetoric." In *Inviting Understanding: A Portrait of Invitational Rhetoric,* edited by Sonja K. Foss and Cindy L. Griffin, 11–16. Lanham, MD: Rowman & Littlefield.

Foss, Sonja K., and Karen A. Foss. 2019. *Inviting Transformation: Presentational Speaking for a Changing World.* Long Grove, IL: Waveland Press.

Foucault, Michel. 1982. *The Archaeology of Knowledge: And the Discourse on Language*. New York: Pantheon Books.

Gaard, Greta. 2017. *Critical Ecofeminism.* Lanham, MD: Lexington Books.

Gallagher, Victoria J., Cynthia Rosenfeld, and Conner Tomlinson. 2022. "Of Sound, Bodies, and Immersive Experience: Sonic Rhetoric and Its Affordances in The Virtual Martin Luther King Project." *enculturation: a journal of rhetoric, writing, and culture*, May 16, 2022. https://www.enculturation.net/Sound_Bodies_ImmersiveExperience

Gallagher, Victoria J., Kelly Norris Martin, and Magdy Ma. 2011. "Visual Wellbeing: Intersections of Rhetorical Theory and Design," *Design Issues* 27, no. 2: 27–40.

Gallagher, Victoria J., and Kenneth S. Zagacki. 2007. "Visibility and Rhetoric: Epiphanies and Transformations in the *Life* Photographs of the Selma Marches of 1965." *Rhetoric Society Quarterly* 37, no. 2: 113–35.

Gallon, Kim. 2016. "Making a Case for the Black Digital Humanities." *Debates in the Digital Humanities*, edited by Matthew K. Gold and Lauren F. Klein, 2016.

https://dhdebates.gc.cuny.edu/read/untitled/section/fa10e2e1-0c3d-4519-a958-d823aac989eb

García, María Elena. 2013. "The Taste of Conquest: Colonialism, Cosmopolitics, and the Dark Side of Peru's Gastronomic Boom." *The Journal of Latin American and Caribbean Anthropology* 18, no. 3: 505–24.

Ghosh, Amitav. 2016. *The Great Derangement: Climate Change and the Unthinkable*. Chicago: University of Chicago Press.

Gilbert, Samuel. 2021. "Native Americans' Farming Practices May Help Feed a Warming World." *Washington Post*, December 10, 2021. https://www.washingtonpost.com/climate-solutions/interactive/2021/native-americans-farming-practices-may-help-feed-warming-world/

Gilligan, Carol. 1993. *In a Different Voice: Psychological Theory and Women's Development.* Cambridge. MA: Harvard University Press.

Ginn, Franklin, Uli Beisel, and Maan Barua. 2014. "Flourishing with Awkward Creatures: Togetherness, Vulnerability, Killing." *Environmental Humanities* 4, no. 1: 113–23.

Giovanni, Nikki. 2021. "Nikki Giovanni Has Made Peace with Her Hate." Interview by David Marchese. *New York Times,* December 26, 2021. https://www.nytimes.com/ interactive/2021/12/27/magazine/nikki-giovanni-interview.html

Giraud, Eva Haifa. 2019. *What Comes After Entanglement? Activism, Anthropocentrism, and an Ethics of Exclusion*. Durham, NC: Duke University Press.

Glusac, Elaine. 2022. "Vegan Travel: It's Not Fringe Anymore." *New York Times*, January 21, 2022. https://www.nytimes.com/2022/01/21/travel/vegan-travel.html

Godfrey-Smith, Peter. 2014. *Philosophy of Biology*. Princeton, NJ: Princeton University Press.

Google. 2022. "Google Keynote." *YouTube*, May 11, 2022. https://youtu.be/nP-nMZpLM1A

Gould, Stephen Jay. 1978. "Sociobiology: The Art of Storytelling." *New Scientist* 80, no. 1129: 530–33.

Graeber, David, and David Wengrow. 2021. *The Dawn of Everything: A New History of Humanity.* New York: Farrar, Straus and Giroux.

Greene, Ronald W. 1998. "Another Materialist Rhetoric." *Critical Studies in Media Communication* 15, no. 1: 21–40.

Greenwalt, Dustin. 2018. "Toward a Rhetorical Ecology." In *Rhetorical Animals: Boundaries of the Human in the Study of Persuasion*, edited by Kristian Bjorkdahl and Alex C. Parrish, 109–28. Lanham, MD: Lexington Books.

Gretzel, Ulrike. 2019. "The Role of Social Media in Creating and Addressing Overtourism." In *Overtourism: Issues, Realities and Solutions*, edited by Rachel Dodds and Richard Butler, 62–75. Boston: De Gruyter.

Gries, Laurie. 2015. *Still Life with Rhetoric: A New Materialist Approach for Visual Rhetoric.* Logan: Utah State University Press.

Gries, Laurie, and Jennifer Clary-Lemon. 2022. "Rhetorical New Materialisms." *Rhetoric Society Quarterly* 52, no. 2: 137–47.

Gries, Laurie E. 2018. "Introduction: Circulation as an Emerging Threshold Concept." In *Circulation, Writing & Rhetoric*, edited by Laurie E. Gries and Collin Gifford Brooke, 3–24. Logan: Utah State University Press.

Grim, John, ed. 2001. *Indigenous Traditions and Ecology: The Interbeing of Cosmology and Community*. Cambridge, MA: Harvard Divinity School.

Gruber, David. 2018. "Multiple Rhetorical Animals: Motivation and Fairness in a Paradigm of Rhetoric as Emotive Consciousness." In *Rhetorical Animals: Boundaries of the Human in the Study of Persuasion*, edited by Kristian Bjorkdahl and Alex C. Parrish, 3–20. Lanham, MD: Lexington Books.

Gruen, Lori. 2015. *Entangled Empathy: An Alternative Ethic for Our Relationships with Animals.* New York: Lantern.

Gudynas, Eduardo. 2011. "*Buen Vivir*: Today's Tomorrow." *Development* 54, no. 4: 441–47.

Haas, Angela. 2015. "Toward a Decolonial Digital and Visual American Indian Rhetorics Pedagogy." In *Survivance, Sovereignty, and Story: Teaching American Indian Rhetorics,* edited by Lisa King, Rose Gubele, and Joyce Rain Anderson, 188–208. Logan: Utah State University Press.

Haas, Angela M. 2007. "Wampum as Hypertext: An American Indian Intellectual Tradition of Multimedia Theory and Practice." *Studies in American Indian Literatures* 19, no. 4: 77–100.

Hancock, Charity, Clifford Hichar, Carlea Holl-Jensen, Kari Kraus, Cameron Mozafari, and Kathryn Skutlin. 2013. "Bibliocircuitry and the Design of the Alien Everyday." *Textual Cultures: Texts, Contexts, Interpretation* 8, no. 1: 72–100.

Haraway, Donna. 1985. "A Cyborg Manifesto: Science, Technology and Socialist-Feminism in the Late Twentieth Century." *Socialist Review* 15, no. 2: 424–57.

———. 1988. "Situated Knowledges: The Science Question in Feminism and the Privilege of Partial Perspective." *Feminist Studies* 14, no. 3: 575–99.

———. 1991. *Simians, Cyborgs, and Women: The Reinvention of Nature*. New York: Routledge.

———. 1998. *How Like a Leaf.* New York: Routledge.

———. 2003. *Companion Species Manifesto: Dogs, People, and Significant Otherness*. Chicago: Prickly Paradigm Press.

———. 2016. *Staying with the Trouble: Making Kin in the Chthulucene.* Durham, NC: Duke University Press.

———. 2018. "Capitalocene and Chthulucene." In *Posthuman Glossary*, edited by Rosi Braidotti and Maria Hlavajova, 79–82. New York: Bloomsbury Academic.

Haraway, Donna, and Anna Tsing. 2019. "Reflections on the Plantationocene," *Edge Effects Magazine*, June 18, 2019. https://edgeeffects.net/wpcontent/uploads/2019/06/ PlantationoceneReflections_Haraway_Tsing.pdf.

Hariman, Robert. 1986. "Status, Marginality, and Rhetorical Theory." *Quarterly Journal of Speech* 72, no. 1: 38–54.

Harper, A. Breeze. 2020. *Sistah Vegan: Black Women Speak on Food, Identity, Health, and Society.* Brooklyn, NY: Lantern Publishing & Media.

Hartelius, Johanna. 2021. "Blockchain and the Virtuous Prospects of Digital Epistemology." Presentation at the National Communication Association 107th Annual Convention, Seattle, WA, November 18–21, 2021.

Haupt, Lyanda Lynn. 2017. *Mozart's Starling*. New York: Back Bay Books.

Hauser, Gerard. 2019. "Attending the Vernacular: A Plea for an Ethnographical Rhetoric." In *Readings in Rhetorical Fieldwork*, edited by Samantha Senda-Cook, Aaron Hess, Michael K. Middleton, and Danielle Endres, 9–20. New York: Routledge.

Haverkamp, Jamie. 2021. "Where's the Love?: Recentering Indigenous and Feminist Ethics of Care for Engaged Climate Research." *Gateways: International Journal of Community Research and Engagement* 14, no. 2: 1–15.

Hawhee, Debra. 2009. *Moving Bodies: Kenneth Burke at the Edges of Language*. Columbia: University of South Carolina Press.

———. 2015. "Rhetoric's Sensorium." *Quarterly Journal of Speech* 101, no. 1: 2–17.

Heidegger, Martin. 1977. *The Question Concerning Technology, and Other Essays*. Translated by William Lovitt. New York: HarperCollins.

Held, Virginia. 2006. *The Ethics of Care: Personal, Political, and Global*. New York: Oxford University Press.

Helweg, David A., Nina Eriksen, Jakob Tougaard, and Lee A. Miller. 2005. "Cultural Change in the Songs of Humpback Whales (*Megaptera novaeangliae*) from Tonga." *Behaviour* 142, no. 3: 305–28.

Herndl, Carl G., Sarah Beth Hopton, Lauren Cutlip, Elena Yu Polushm, Rick Cruse, and Mack Shelley. 2018. "What's a Farm? The Languages of Space and Place." In *Field Rhetoric: Ethnography, Ecology, and Engagement in the Places of Persuasion,* edited by Candice Rai and Caroline Gottschalk Druschke, 61–94. Tuscaloosa: University of Alabama Press.

Herzog, Hal. 2010. *Some We Love, Some We Hate, Some We Eat.* New York: Harper Perennial.

Hidalgo-Capitán, Antonio Luis, Ana Patricia Cubillo-Guevara, and Francisco Masabalín-Caisaguano. 2020. "The Ecuadorian Indigenist School of Good Living (*Sumak Kawsay*)." *Ethnicities* 20, no. 3: 408–33.

Hird, Myra J. 2016. "The Phenomenon of Waste-World-Making." *Rhizomes: Cultural Studies in Emerging Knowledge* 30: 183–91.

Holland, Jennifer S. 2015. "Quokka Selfies: What's the Deal With That Cute Australian Critter?" *National Geographic*, March 6, 2015. https://www.nationalgraphic.com/animals/article/150306-quokkas-selfies-animals-science-photography-australia

Hu, Frank B., Brett O. Otis, and Gina McCarthy. 2019. "Can Plant-Based Meat Alternatives Be Part of a Healthy and Sustainable Diet?" *Journal of the American Medical Association* 322, no. 16: 1547–48.

Hughes, Neil. 2021. "Exploring Vegan Ideology through Graffiti Slogans." 2021. *Discourse & Society* 32, no. 5: 575–97.

Imarisha, Walidah. 2015. "Introduction." In *Octavia's Brood: Science Fiction Stories from Social Justice Movements*, edited by Adrienne Maree Brown and Walidah Imarisha, 3–5. Oakland, CA: AK Press.

Impossible Foods. "Mission." *Impossible*. Accessed July 1, 2022. https://impossiblefoods.com/sustainable-food

Ingraham, Chris. 2018. “Energy: Rhetoric’s Vitality.” *Rhetoric Society Quarterly* 48, no. 3: 260–68.

———. 2020. *Gestures of Concern.* Durham, NC: Duke University Press.

Inoue, Asao B. 2019. *Labor-Based Grading Contracts: Building Equity and Inclusion in the Compassionate Writing Classroom*. Fort Collins, CO: WAC Clearinghouse.

Isbell, Lynne A. 2009. *The Fruit, the Tree, and the Serpent: Why We See Well.* Cambridge, MA: Harvard University Press.

Isbell, Lynne A., and Stephanie F. Etting. 2017. “Scales Drive Detection, Attention, and Memory of Snakes in Wild Vervet Monkeys (*Chlorocebus Pygerythrus*).” *Primates* 58, no. 1: 121–29.

Jackson, Gerald. 2018. “Crytpocurrency and Persuasive Network Logics: From the Circulation of Rhetoric to the Rhetoric of Circulation.” In *Circulation, Writing & Rhetoric*, edited by Laurie E. Gries and Collin Gifford Brooke, 225–42. Logan: Utah State University Press.

Jackson, Sarah J., Moya Bailey, and Brooke Foucault Welles. 2020. *#Hashtag Activism: Networks of Race and Gender Justice.* Cambridge, MA: MIT Press

Jackson, Scott, and Peter Mirick. *Massachusetts Snakes: A Guide* (University of Massachusetts Extension), accessed June 29, 2022. https://ag.umass.edu/sites/ag.umass.edu/files/pdf-doc-ppt/ma_snake_guide.pdf

Jasinski, James. 2001. *Sourcebook on Rhetoric.* Thousand Oaks, CA: Sage.

Jensen, Derek. 2016. *The Myth of Human Supremacy.* New York: Seven Stories Press.

Johnson, Mark. 2007. *The Meaning of the Body: Aesthetics and Human Understanding.* Chicago: University of Chicago Press.

Johnson, Steven. 2011. *Where Good Ideas Come From*. New York: Riverhead Books.

Jørgensen, Charlotte. 2007. “The Relevance of Intention in Argument Evaluation.” *Argumentation* 21, no. 2: 165–74.

Josephson, Paul R. 2020. *Chicken: A History from Farmyard to Factory*. Medford, MA: Polity Press.

Jue, Melody. 2020. *Wild Blue: Thinking Through Seawater*. Durham, NC: Duke University Press.

Karasti, Helena, Volkmar Pipek, and Geoffrey C. Bowker. 2018. “An Afterword to ‘Infrastructuring and Collaborative Design.’” *Computer Supported Cooperative Work* 27, no. 2: 267–89.

Kasturiratne, Anuradhani, A. Rajitha Wickremasinghe, Nilanthi de Silva, N. Kithsiri Gunawardena, Arunasalam Pathmeswaran, Ranjan Premaratna, Lorenzo Savioli, David G. Lalloo, and H. Janaka de Silva. 2008. “The Global Burden of Snakebite: A Literature Analysis and Modelling Based on Regional Estimates of Envenoming and Deaths.” *PLoS Medicine* 5, no. 11: 1591–1604.

Kemmerer, Lisa. 2017. “Indigenous Traditions.” In *Food, Ethics, and Society: An Introductory Test with Readings*, edited by Anne Barnhill, Mark Budolfson, and Tyler Doggett, 309–21. New York: Oxford University Press.

Kennedy, George A. 1992. “A Hoot in the Dark: The Evolution of General Rhetoric.” *Philosophy & Rhetoric* 25, no. 1: 1–21.

Kennedy, George. 1998. *Comparative Rhetoric: An Historical and Cross-Cultural Introduction*. New York: Oxford University Press.

Kennerly, Michele, and Damien Smith Pfister, eds. 2018. *Ancient Rhetorics and Digital Networks*. Tuscaloosa: University of Alabama Press.

Keysers, Christian, and Valeria Gazzola. 2014. "Hebbian Learning and Predictive Mirror Neurons for Actions, Sensations and Emotions." *Philosophical Transactions of the Royal Society B: Biological Sciences* 369, no. 1644: 1–11.

Kim, Claire Jean. 2015. *Dangerous Crossings: Race, Species, and Nature in a Multicultural Age*. New York: Cambridge University Press.

———. 2018. "Abolition." In *Critical Terms for Animal Studies*, edited by Lori Gruen, 15–32. Chicago: University of Chicago Press.

Kimmerer, Robin Wall. 2013. *Braiding Sweetgrass: Indigenous Wisdom, Scientific Knowledge, and the Teachings of Plants.* Minneapolis, MN: Milkweed Editions.

Kingsford, Anna Bonus. 2018. *The Perfect Way in Diet*. London: Anna Bonus Kingsford.

Kirsch, Gesa E. 2013. "A Land Ethic for Urban Dwellers." In *Environmental Rhetoric and Ecologies of Place,* edited by Peter N. Goggin, 69–83. New York: Routledge.

Kleiman, Devra G., and John Seidensticker. 1985. "Pandas in the Wild." *Science* 228, no. 4701: 875–76.

Klein, Naomi. 1999. *No Logo.* New York: Picador.

———. 2014. *This Changes Everything: Capitalism vs. the Climate.* New York: Simon & Schuster.

Knutson, Ted. 2022. "Crypto Increasingly Used in Human/Drug Trafficking Says GAO." *Forbes*, January 10, 2022. https://www.forbes.com/sites/tedknutson/2022/01/10/crypto-increasingly-used-in-humandrug-trafficking-says-gao/?sh=2c1586c9637e

Ko, Syl, and Aph Ko. 2017. *Aphro-ism: Essays on Pop Culture, Feminism, and Black Veganism from Two Sisters*. Brooklyn, NY: Lantern Publishing & Media.

Kohn, Eduardo. 2013. *How Forests Think: Toward an Anthropology Beyond the Human*. Berkeley: University of California Press.

Kolbert, Elizabeth. 2014. *The Sixth Extinction: An Unnatural History*. New York: Picador.

———. 2021. *Under a White Sky: The Nature of the Future*. New York: Crown.

Korsgaard, Christine. 2018. *Fellow Creatures: Our Obligations to the Other Animals.* New York: Oxford University Press.

Kreilkamp, Ivan. 2020. "How Caring for Backyard Chickens Stretched My Emotional Muscles." *New York Times,* November 25, 2020. https://www.nytimes.com/2020/11/25/magazine/backyard-chickens-empathy.html

Kress, Gunther, and Theo van Leeuwen. 2021. *Reading Images: The Grammar of Visual Design*. New York: Routledge.

Kress, W. John, and Jeffrey K. Stein, eds. 2017. *Living in the Anthropocene: Earth in the Age of Humans.* New York City: Soho Press.

LaDuke, Winona. 2020. *To Be a Water Protector: The Rise of the Wiindigoo Slayers.* Winnipeg: Fernwood Publishing, 2020.

Lakoff, George, and Mark Johnson. 2003. *Metaphors We Live By*. Chicago: University of Chicago Press.

Larsson, Agnes. 2022. "The Awesome Potential of Many Metaverses." *TED2022*. https://www.ted.com/talks/agnes_larsson_the_awesome_potential_of_many_metaverses?language=en

Latour, Bruno. 2000. "Opening Pandora's Black Box." In *Technology, Organizations, and Innovation: Critical Perspectives on Business and Management*, edited by David Preece, Ian McLoughlin, and Patrick Dawson, 679–97. New York: Routledge.

———. 2007. *Reassembling the Social: An Introduction to Actor-Network Theory*. New York: Oxford University Press.

Lecointre, Guillaume, and Hervé Le Guyader. 2006. *The Tree of Life: A Phylogenetic Classification*. Cambridge, MA: The Belknap Press.

Ledbetter, Andrew M., Joseph P. Mazer, Jocelyn M. DeGroot, Kevin R. Meyer, Yup-ing Mao, and Brian Swafford. 2011. "Attitudes Toward Online Social Connection and Self-Disclosure as Predictors of Facebook Communication and Relational Closeness." *Communication Research* 38, no. 1: 27–53.

Ledur, Julia, and Kate Rabinowitz. 2022. "There Have Been over 250 Mass Shootings So Far in 2022." *Washington Post*, June 2, 2022. https://www.washingtonpost.com/nation/2022/06/02/mass-shootings-in-2022/

Leff, Michael C. 1980. "Interpretation and the Art of the Rhetorical Critic." *Western Journal of Speech Communication* 44, no. 4: 337–49.

Leff, Michael. 1996. "Commonplaces and Argumentation in Cicero and Quintilian." *Argumentation* 10, no. 4: 445–452.

Leonard, Annie. 2010. *The Story of Stuff*. New York: Free Press.

Leopold, Aldo. 1987. *A Sand County Almanac: And Sketches Here and There*. New York: Oxford University Press.

Levin, Ted. 2016. *America's Snake: The Rise and Fall of the Timber Rattlesnake*. Chicago: University of Chicago Press.

Lewis, Simon, and Mark Maslin. 2018. *The Human Planet: How We Created the Anthropocene*. New Haven, CT: Yale University Press.

Lewis, Tania. 2020. *Digital Food: From Paddock to Platform*. New York: Bloomsbury.

Lieberman, Matthew D. 2013. *Social: Why Our Brains Are Wired to Connect*. New York: Crown Publishers.

Lin, Brenda B., M. Jahi Chappell, John Vandermeer, Gerald Smith, Eileen Quintero, Rachel Bezner-Kerr, and Daniel M. Griffith et al. 2011. "Effects of Industrial Agriculture on Climate Change and the Mitigation Potential of Small-Scale Agro-ecological Farms." *CABI Reviews* 6, no. 20: 1–18.

LoBue, Vanessa. 2014. "Deconstructing the Snake: The Relative Roles of Perception, Cognition, and Emotion on Threat Detection." *Emotion* 14, no. 4: 701–11.

LoBue, Vanessa, and Karen E. Adolph. 2019. "Fear in Infancy: Lessons from Snakes, Spiders, Heights, and Strangers." *Developmental Psychology* 55, no. 9: 1889–1907.

Lozano, Nina Maria. 2019. *Not One More! Femincido on the Border*. Columbus: Ohio State University Press.

Lynch, Amanda H., and Siri Veland. 2018. *Urgency in the Anthropocene*. Cambridge, MA: MIT Press.

Macdonald, Helen. 2015. *H Is for Hawk*. New York: Grove Press.

MacKinnon, James B. 2018. "Tragedy of the Common." *Pacific Standard,* September 15, 2018. https://psmag.com/magazine/tragedy-of-the-common

Mann, Michael E. 2021. *The New Climate War: The Fight to Take Back Our Planet.* New York: PublicAffairs.

Mann, Michael E., and Lee R. Kump. 2015. *Dire Predictions: Understanding Climate Change*. New York: DK Publishing.

Marchesini, Roberto. 2015. "Against Anthropocentrism: Non-Human Otherness and the Post-Human Project." *NanoEthics* 9, no. 1: 75–84.

Marcum, Heidi. 2007. "Living with Animals: Snakes and Humans" in *Encyclopedia of Human-Animal Relationships*, edited by Marc Bekoff, 1181–84. Westport, CT: Greenwood Press.

Marris, Emma. 2021. *Wild Souls: Freedom and Flourishing in the Non-human World.* New York: Bloomsbury Publishing.

Marx, Karl. 1972. "The Eighteenth Brumaire of Louis Bonaparte." In *The Karl Marx Library, Vol. 1*, edited by Saul K. Padover, 245–46. New York: McGraw Hill.

Massumi, Brian. 2002. *Parables for the Virtual: Movement, Affect, Sensation.* Durham, NC: Duke University Press.

McHugh, Susan. 2011. *Animal Stories: Narrating Across Species Lines*. Minneapolis: University of Minnesota Press.

McKinnon, Sara L., Robert Asen, Karma R. Chávez, and Robert Glenn Howard, eds. 2016. *Text+ Field: Innovations in Rhetorical Method.* University Park: Penn State Press.

McQuirter, Tracye. 2021. "Call to Action: It's about Your Greens, Not Your Genes." In *Black Food*, edited by Bryant Terry, 170–71. New York: 4 Color Books.

Meek, Laura A., and Julia Alejandra Morales Fontanilla. 2022. "Otherwise." *Feminist Anthropology*, June 4, 2022. https://doi.org/10.1002/fea2.12094

Meighan, Paul J. 2021. "Decolonizing English: A Proposal for Implementing Alternative Ways of Knowing and Being in Education." *Diaspora, Indigenous, and Minority Education* 15, no. 2: 77–83.

Meta. 2022. Terms of Service. *Facebook.* https://www.facebook.com/terms.php?ref=pf

Middha, Bhavna. 2018. "Everyday Digital Engagements: Using Food Selfies on Facebook to Explore Eating Practices." *Communication Research and Practice* 4, no. 3: 291–306.

Miller, L., P. Rozin, and A. P. Fiske (1998). "Food Sharing and Feeding Another Person Suggest Intimacy: Two Studies of American College Students." *European Journal of Social Psychology* 28, no. 3: 423–36.

Miller, William R., and Stephen Rollnick. 2012. *Motivational Interviewing: Helping People Change*, 3rd ed. New York: The Guilford Press.

Milstein, Tema. 2011. "Nature Identification: The Power of Pointing and Naming." *Environmental Communication* 5, no. 1: 3–24.

Milstein, Tema, and Charlotte Kroløkke. 2012 "Transcorporeal Tourism: Whales, Fetuses, and the Rupturing and Reinscribing of Cultural Constraints." *Environmental Communication* 6, no. 1: 82–100.

Milstein, Tema, and José Castro-Sotomayor. 2020. "Ecocultural Identity: An Introduction." In *Routledge Handbook of Ecocultural Identity*, edited by Tema Milstein and José Castro-Sotomayor, xvii–xxiii. New York: Routledge.

Mineka, Susan, and Michael Cook. 1986. "Immunization Against the Observational Conditioning of Snake Fear in Rhesus Monkeys." *Journal of Abnormal Psychology* 95, no. 4: 307–18.

Mining[Dot]Com. 2022. "Western Australia Is World's New Top Mining Destination." *Mining[Dot]Com*, April 11, 2022. https://www.mining.com/western-australia-is-worlds-new-top-mining-destination/

Miyarrka Media. 2019. *Phone & Spear: A Yuta Anthropology.* London: Goldsmiths Press.

Montgomery, Sy. 2022. *The Hawk's Way: Encounters with Fierce Beauty*. New York: Atria Books.

Moore, Peter. 2014. "Argh! Snakes! America's Top Phobias Revealed." *YouGovAmerica,* March 27, 2014. https://today.yougov.com/topics/lifestyle/articles-reports/2014/03/27/argh-snakes

Morton, Timothy. 2013. *Hyperobjects: Philosophy and Ecology after the End of the World*. Minneapolis: University of Minnesota Press.

Muckelbauer, John. 2021. "Asignification." In *A New Handbook of Rhetoric: Inventing the Classical Vocabulary*, edited by Michele Kennerly, 38–49. University Park: Penn State University Press.

Mulvey, Laura. 1989. *Visual and Other Pleasures.* New York: Palgrave Macmillan.

Mwangi, Evan Maina. 2019. *The Postcolonial Animal: African Literature and Posthuman Ethics.* Ann Arbor: University of Michigan Press.

Myrick, Jessica Gall, and Suzannah D. Evans. 2014. "Do PSAs Take a Bite Out of Shark Week? The Effects of Juxtaposing Environmental Messages with Violent Images of Shark Attacks." *Science Communication* 36, no. 5: 544–69.

National Chicken Council. 2021. "Per Capita Consumption of Poultry and Livestock, 1965 to Forecast 2022, in Pounds." National Chicken Council. https://www.nationalchickencouncil.org/about-the-industry/statistics/per-capita-consumption-of-poultry-and-livestock-1965-to-estimated-2012-in-pounds/

Native Governance Center. 2019. "A Guide to Indigenous Land Acknowledgement." *Native Governance Center.* https://nativegov.org/news/a-guide-to-indigenous-land-acknowledgment/

Native Languages of the Americas. 2020. "Native American Rattlesnake Mythology." *Native Languages of the Americas: Preserving and Promoting American Indian Languages*. http://www.native-languages.org/legends-rattlesnake.htm

Newberry, Christina. 2022. "How the Facebook Algorithm Works in 2023 and How to Make It Work for You." *Hootsuite*, February 22, 2022. https://blog.hootsuite.com/facebook-algorithm/#:~:text=The%20Facebook%20algorithm%20determines%20which,interest%20for%20each%20individual%20user

Nisbet, Jeff. 2019. *This Is the Way the World Ends: How Droughts and Die-offs, Heat Waves and Hurricanes Are Converging on America.* New York: Picador.

Noble, Safiya. 2018. *Algorithms of Oppression: How Search Engines Reinforce Racism.* New York: New York University Press.

Noisy Hawk, Lyle J., and Joseph E. Trimble. 2019. "Wellbeing Considerations among Selected North American Indian Populations: Relationships, Spirits and Connections." In *Routledge Handbook of Indigenous Wellbeing*, edited by Christoper Fleming and Matthew Manning, 97–108. New York: Routledge.

Nussbaum, Martha. 2006a. *Frontiers of Justice: Disability, Nationality, Species Membership.* Cambridge, MA: The Belknap Press.

———. 2006b. "The Moral Status of Animals." *Chronicle of Higher Education* 52, no. 22. https://www.chronicle.com/article/the-moral-status-of-animals/

———. 2022. "What We Owe Our Fellow Animals." *The New York Review of Books* LXIX, no. 4. https://www.nybooks.com/articles/2022/03/10/what-we-owe-our-fellow-animals-ethics-martha-nussbaum/

O'Gorman, Marcel. 2019. "Making Media Theory with a Smartphone Basket." *enculturation: a journal of rhetoric, writing, and culture* 29: enculturation.net/a_smartphone_basket

O'Neil, Cathy. 2016. *Weapons of Math Destruction*. New York: Crown Publishers.

Ohlheiser, Abby. "The Woman Behind 'Me Too' Knew the Power of the Phrase When She Created It—10 Years Ago." *Washington Post,* October 19, 2017. https://www.washingtonpost.com/news/the-intersect/wp/2017/10/19/the-woman-behind-me-too-knew-the-power-of-the-phrase-when-she-created-it-10-years-ago/

Öhman, Arne, and Susan Mineka. 2001. "Fears, Phobias, and Preparedness: Toward an Evolved Module of Fear and Fear Learning." *Psychological Review* 108, no. 3: 483–522.

Omnicore. 2022. "Instagram by the Numbers: Stats, Demographics, & Fun Facts." *Omnicore,* February 27, 2022. https://www.omnicoreagency.com/instagram-statistics/

Opperman, Serpil. 2014. "From Ecological Postmodernism to Material Ecocriticism: Creative Materiality and Narrative Agency." In *Material Ecocriticism*, edited by Serenella Iovino and Serpil Opperman, 21–36. Bloomington: Indiana University Press.

Pal, Sanjoy Kumar, Aparna Gomes, S. C. Dasgupta, and Antony Gomes. 2002. "Snake Venom as Therapeutic Agents: From Toxin to Drug Development." *Indian Journal of Experimental Biology* 40, no. 12: 1353–58.

Parks, Lisa, and Nicole Starosielski. 2015. "Introduction." In *Signal Traffic: Critical Studies of Media Infrastructures*, edited by Lisa Parks and Nicole Starosielski, 1–27. Urbana: University of Illinois Press.

Parrish, Alex. 2014. *Adaptive Rhetoric: Evolution, Culture, and the Art of Persuasion.* New York: Routledge.

Pauly, Daniel. 2014. "Homo Sapiens: Cancer or Parasite?" *Ethics in Science and Environmental Politics* 14, no. 1: 7–10.

Paxman, Christina G. 2021. "'I Love Tater Tot Casserole, I Just Make It Vegan': Applying the Communication Theory of Identity to Examine Vegans' Identity Management Techniques." *Communication Studies* 72, no. 4: 752–68.

Peredo, Ana Maria. 2019. "*El Buen Vivir*: Notions of Wellbeing among Indigenous Peoples of South America." In *Routledge Handbook of Indigenous Wellbeing,* edited by Christoper Fleming and Matthew Manning, 156–69. New York: Routledge.

Peirce, Charles Sanders. 1977. *Semiotics and Significs*, edited by Charles Hardwick. Bloomington: Indiana University Press.

———. 1992. "Evolutionary Love (1893)." In *The Essential Peirce: Selected Philosophical Writings Volume 1 (1867–1893)*, edited by Nathan Houser and Christian Kloesel, 352–71. Bloomington: Indiana University Press.

———. 1998. *The Essential Peirce: Selected Philosophical Writings Volume 2 (1893–1913)*, edited by the Peirce Edition Project. Bloomington: Indiana University Press.

Perkins, Devin N., Marie-Noel Brune Drisse, Tapiwa Nxele, and Peter D. Sly. 2014. "E-waste: A Global Hazard." *Annals of Global Health* 80, no. 4: 286–95.

Peters, John Durham. 2015. *The Marvelous Clouds*. Chicago: University of Chicago Press.

Peters, John Durham, Florian Sprenger, and Christina Vagt. 2020. *Action at a Distance*. Minneapolis: University of Minnesota Press.

Petronio, Sandra. 2013. "Brief Status Report on Communication Privacy Management Theory." *Journal of Family Communication* 13, no. 1: 6–14.

Pezzullo, Phaedra, and Robert Cox. 2021. *Environmental Communication and the Public Sphere*, 6th ed. Thousand Oaks, CA: Sage.

Pezzullo, Phaedra C., and Catalina M. de Onís. 2018. "Rethinking Rhetorical Field Methods on a Precarious Planet." *Communication Monographs* 85, no. 1: 103–22.

Phelps, Norm. 2007. *The Longest Struggle: Animal Advocacy from Pythagoras to PETA*. New York: Lantern.

———. 2016. "Digital Ethnography." In *Innovative Methods in Media and Communication Research*, edited by Sebastian Kubitschko and Anne Kaun, 161–65. New York: Palgrave Macmillan.

Pink, Sarah, Heather Horst, John Postill, Larissa Hjorth, Tania Lewis, and Jo Tacchi. 2015. *Digital Ethnography: Principles and Practice*. Thousand Oaks, CA: Sage.

Plato. 2001. "Gorgias." In *The Rhetorical Tradition: Readings from Classical Times to Present,* edited by Patricia Bizzell and Bruce Herzberg, 87–137. Boston: Bedford/St. Martin's.

Plumwood, Val. 2002. *Environmental Culture: The Ecological Crisis of Reason.* New York: Routledge.

———. 2012. *The Eye of the Crocodile*, edited by Lorraine Shannon. Canberra: ANU E Press.

Pogue, David. 2021. *How to Prepare for Climate Change: A Practical Guide to Surviving the Chaos.* New York: Simon & Schuster.

Poole, Megan. 2020. "Orientation: Seeing and Sensing Rhetorically." *Western Journal of Communication* 84, no. 5: 604–22.

Postill, John. 2017. "Remote Ethnography: Studying Culture from Afar." In *The Routledge Companion to Digital Ethnography,* edited by Larissa Hjorth, Heather Horst, Anne Galloway, and Genevieve Bell, 61–69. New York: Routledge.

Potts, Annie. 2012. *Chicken*. London: Reaktion Books.

Prabasmoro, Tisna, Rasus Budhyono, and Teddi Muhtadin. 2019. "Black Panther: Identity, Afrofuturism, and Representation of Superheroes." *Pertanika Journal of Social Sciences & Humanities* 27, no. 4: 2672–85.

Prelli, Lawrence J. 2006. *Rhetorics of Display.* Columbia: University of South Carolina Press.

Pritchford, Matt. 2022. "Are Memes Topoi?: Digital Cultures and the Function, Form, and Content of *Topoi* Online." Presentation at the 20th Biennial Conference of the Rhetoric Society of America, Baltimore, MD, May 26–29, 2022.

Prum, Richard. 2017. *The Evolution of Beauty: How Darwin's Forgotten Theory of Mate Choice Shapes the Animal World—and Us.* New York: Anchor Books.

Quartz, Sean. 2022. "Becoming More-Than-Human: Realizing Earthly Eudaimonia to (E)coflourish through an Entangled Ethos." *Journalism and Media* 3, no. 2: 238–53.

Ranteallo, I. C., and I. R. Andilolo. 2017. "Food Representation and Media: Experiencing Culinary Tourism through Foodgasm and Foodporn," in *Balancing Development and Sustainability in Tourism Destinations*, 117–27. Singapore: Springer.

Ratto, Matt. 2011. "Critical Making: Conceptual and Material Studies in Technology and Social Life." *The Information Society* 27, no. 4: 252–60.

Ratto, Matt, and Stephen Hockema. 2009. "Flwr Pwr: Tending the Walled Garden." *Critical Making Lab.* https://criticalmaking.com/wp-content/uploads/2009/10/2448_alledgarden_ch06_ratto_hockema.pdf

Reading, Richard. 2005. "The Enduring Effects of Abuse and Related Adverse Experiences in Childhood: A Convergence of Evidence from Neurobiology and Epidemiology." *European Archives of Psychiatry and Clinical Neuroscience*, November 29, 2005. doi: 10.1111/j.1365-2214.2006.00614_2.x

Reber, Paul. 2010. "What Is the Memory Capacity of the Human Brain?" *Scientific American,* May 1, 2010. https://www.scientificamerican.com/article/what-is-the-memory-capacity/#:~:text=You%20might%20have%20only%20a,(or%20a%20million%20gigabytes)

Reclaim the Void. *Reclaim the Void: Weaving Country Whole*, accessed July 2, 2022. https://www.reclaimthevoid.com.au/

Regan, Tom. 2004. *The Case for Animal Rights*. Berkeley: University of California Press.

Renner, Max Michael. 2020. "Building Architects, Inventing Publics: Re-mapping the Lifespan of Public Places." PhD diss., North Carolina State University.

Rich, Nathaniel. 2019. *Losing Earth: A Recent History*. New York: MCD/Farrar, Straus and Giroux.

Rickert, Thomas. 2013. *Ambient Rhetoric: The Attunements of Rhetorical Being*. Pittsburgh: University of Pittsburgh Press.

Riley, Matthew. 2014. "*A Spiritual Democracy of All God's Creatures: Ecotheology and the Animals of Lynn White Jr.*" In *Divinanimality: Animal Theory, Creaturely Theology*, edited by Stephen Moore, 241–60. New York: Fordham University Press.

Risam, Roopika. 2019. *New Digital Worlds: Postcolonial Digital Humanities in Theory, Praxis, and Pedagogy.* Evanston, IL: Northwestern University Press.

Roberts, James A., and Meredith E. David. 2020. "The Social Media Party: Fear of Missing Out (FoMO), Social Media Intensity, Connection, and Well-being." *International Journal of Human–Computer Interaction* 36, no. 4: 386–92.

Rodrigues, Caroline. 2022. “Investment Opportunity Breakdown: Plant-Based Food Stocks.” *Nasdaq*, January 14, 2022. https://www.nasdaq.com/articles/investment-opportunity-breakdown%3A-plant-based-food-stocks

Romo, Lynsey Kluever, and Erin Donovan-Kicken. 2012. “‘Actually, I Don’t Eat Meat’: A Multiple-Goals Perspective of Communication about Vegetarianism.” *Communication Studies* 63, no. 4: 405–20.

Roose, Kevin. 2023. “A.I. Poses ‘Risk of Extinction,’ Industry Leaders Warn.” *New York Times*, May 30, 2023. https://www.nytimes.com/2023/05/30/technology/ai-threat-warning.html

Rose, Deborah Bird. 1988. “Exploring an Aboriginal Land Ethic.” *Meanjin* 47, no. 3: 378–87.

Rosenfeld, Cynthia. 2019. “From Prometheus to Gaea: A Case for Earth-Centered Language.” *Language & Ecology* 2019: 1–16.

———. 2020. “‘Oh! They’re Not Slimy!’ An Ethnographic Exploration of Human-Snake Encounters.” *Journal for Critical Animal Studies* 17, no. 2: 13–41.

———. 2021. “Bowers of Persuasion: Toward a Posthuman Visual Rhetoric.” In *Posthumanist Perspectives on Literary and Cultural Animals*, edited by Krishanu Maiti, 75–86. New York: Springer.

———. 2021. “Slithering Stories We Live By: Animal Educators’ Construction and Enactment of Positive Snake Narratives.” *Society & Animals* 2, no. 10: 1–18.

Rosenfeld, Lawrence B. 1979. “Self-Disclosure Avoidance: Why I Am Afraid to Tell You Who I Am.” *Communications Monographs* 46, no. 1: 63–74.

Roth, Geneen. 1992. *When Food Is Love: Exploring the Relationship between Eating and Intimacy*. London: Penguin.

Sackey, Donnie. 2022. “From Precarity to Invasion: Nonhuman Rhetorics in Times of Global Climate Change.” Presentation at the 20th Biennial Conference of the Rhetoric Society of America, Baltimore, MD, May 26–29, 2022.

Saey, Tina Hesman. 2022. “A Weapon Against Mosquitoes.” *Science News*, June 4, 2022, 20–25.

Scheyder, Ernest. 2022. “U.S. Court Upholds Arizona Land Swap Deal for Rio Tinto Copper Mine.” *Reuters*, June 27, 2022. https://www.reuters.com/world/us/us-court-upholds-arizona-land-swap-deal-rio-tinto-copper-mine-2022-06-25/

Schipani, Sam. 2018. “At Rattlesnake Roundups, Cruelty Draws Crowds.” *Sierra,* March 10, 2018. https://www.sierraclub.org/sierra/rattlesnake-roundups-sweetwater-texas-cruelty

Scranton, Roy. 2015. *Learning to Die in the Anthropocene.* San Francisco: City Lights Publishers.

Seal, Ananya, SubbaRao M. Gavaravarapu, and Archana Konapur. 2022. “Can Foodporn Prime Healthy Eating? Thinking Beyond Digital Gazing and Satiety.” *European Journal of Clinical Nutrition* 76, no. 11: 1–4.

Seegert, Natasha. 2014. “Play of Sniffication: Coyotes Sing in the Margins.” *Philosophy & Rhetoric* 47, no. 2: 158–78.

———. 2016. “Rewilding Rhetoric with Animate Others.” *Review of Communication* 16, no. 1: 77–79.

Senda-Cook, Samantha, and George F. McHendry Jr. 2018. "Embodying Resistance: A Rhetorical Ecology of the Full Cycle Supper." In *Tracing Rhetoric and Material Life,* edited by Bridie McGreavy, Justine Wells, George F. McHendry Jr., and Samantha Senda-Cook, 223–52. New York: Palgrave Macmillan.

Senda-Cook, Samantha, Aaron Hess, Michael K. Middleton, and Danielle Endres, eds. 2018. *Readings in Rhetorical Fieldwork*. New York: Routledge.

Seraphin, Bruno. 2020. "Wildtending, Settler Colonialism, and Ecocultural Identities in Environmental Futures." In *Routledge Handbook of Ecocultural Identity,* edited by Tema Milstein and José Castro-Sotomayor, 403–15. New York: Routledge.

Settee, Priscilla. 2018. "Indigenous Food Sovereignty in Canada." In *Traditional Ecological Knowledge: Learning from Indigenous Practices for Environmental Sustainability*. Cambridge: Cambridge University Press.

Sharma, Sarah. 2017. "Exit and Extensions of Man." *Transmediale/Art and Digital Culture*, August 5, 2017. https://transmediale.de/content/exit-and-the-extensions -of-man.

Shedd, John. 1928. *Salt from My Attic*. Portland, ME: Mosher Press.

Shellenberger, Michael, and Ted Nordhaus. 2011. "Evolve." *Orion.* https:// orionmagazine.org/article/evolve/

Simonsen, Jesper, Helena Karasti, and Morten Hertzum. 2020. "Infrastructuring and Participatory Design: Exploring Infrastructural Inversion as Analytic, Empirical and Generative." *Computer Supported Cooperative Work* 29, no. 1: 115–51.

Singer, Peter. 2009. *Animal Liberation*. New York: Harper Perennial.

Smith, Kit. 2020. "60 Incredible and Interesting Twitter Stats and Statistics." *Brandwatch,* January 2, 2020. https://www.brandwatch.com/blog/twitter-stats-and -statistics/#:~:text=Twitter%20usage %20statistics,That's%206%2C000%20tweets %20every%20second

Soledad del Villar, María. 2018. "The Küme Mongen and the Reign of God." *Lumen et Vita* 9, no. 1.

Spinoza, Benedict de. 1996. *Ethics.* Translated by Edwin Curley. New York: Penguin Books.

Springer, Michael, and Peter N. Goggin. 2013. "Digital Cities: Rhetorics of Place in Environmental Video Games." In *Environmental Rhetoric and Ecologies of Place,* edited by Peter N. Goggin, 111–26. New York: Routledge.

Star, Susan Leigh. 1999. "The Ethnography of Infrastructure." *American Behavioral Scientist* 43, no. 3: 377–91.

Star, Susan Leigh, and Karen Ruhleder. 1996. "Steps Toward an Ecology of Infrastructure: Design and Access for Large Information Spaces." *Information Systems Research* 7, no. 1: 111–34.

Statista. 2022. "Most Popular Social Networks Worldwide as of October 21, Ranked by Number of Active Users." *Statista*, January 28, 2022. https://www.statista.com/ statistics/272014/global-social-networks-ranked-by-number-of-users/

Steinberger, Michael. 2022. "Novak Djokovic, a Master on the Court, Keeps Making Errors Off It." *New York Times,* January 16, 2022, https://www.nytimes.com/2022 /01/15/ sports/tennis/novak-djokovic-australian-open.html

Stibbe, Arran. 2018. "Positive Discourse Analysis: Rethinking Human and Ecological Relationships." In *Routledge Handbook of Ecolinguistics,* edited by Arran Stibbe and Hermine Penz, 165–78. New York: Routledge.

———. 2021. *Ecolinguistics: Language, Ecology and the Stories We Live By.* New York: Routledge.

Strubenhoff, Heinz. 2022. "The War in Ukraine Triggered a Global Food Shortage." *Brookings*, June 14, 2022. https://www.brookings.edu/blog/future-development/2022/06/14/the-war-in-ukraine-triggered-a-global-food-shortage/

Taylor, Angus. 2003. *Animals & Ethics.* Orchard Park: Broadview Press.

Taylor, Sunaura. 2017. *Beasts of Burden: Animal and Disability Liberation.* New York: The New Press.

Tessman, Lisa. 2005. *Burdened Virtues: Virtue Ethics for Liberatory Struggles.* New York: Oxford University Press.

The Chamber of Minerals and Energy in Western Australia. 2022. "Map of Resource Projects in Western Australia." *The Chamber of Minerals and Energy in Western Australia.* https://www.cmewa.com.au/about/wa-resources/project-map/

The Pink Chicken Project. *The Pink Chicken Project*, accessed July 2, 2022. https://pinkchickenproject.com/

Thunberg, Greta. 2019. *No One Is Too Small to Make a Difference.* New York: Penguin Books.

Todd, Zoe. 2015. "Indigenizing the Anthropocene." In *Art in the Anthropocene: Encounters among Aesthetics, Politics, Environments and Epistemologies*, edited by Heather Davis and Etienne Turpin, 241–54. London: Open Humanities Press.

Towns, Armond R. 2022. *On Black Media Philosophy*. Oakland: University of California Press.

Tronto, Joan. 1993. *Moral Boundaries: A Political Argument for an Ethic of Care.* New York: Routledge.

Tsing, Anna. 2015. *The Mushroom at the End of the World: On the Possibility of Life in Capitalist Ruins.* Princeton, NJ: Princeton University Press.

University of Minnesota. 2022. "Accessibility vs. Accommodation." *Accessible U.* https://accessibility.umn.edu/importance-accessibility/accessibility-vs-accommodation

Valladolid, Julio, and Frédérique Apffel-Marglin. 2001. "Andean Cosmovision and the Nurturing of Biodiversity." In *Indigenous Traditions and Ecology: The Interbeing of Cosmology and Community*, edited by John Grim, 639–70. Cambridge, MA: Harvard Divinity School.

van Dooren, Thom. 2014. *Flight Ways: Life and Loss at the Edge of Extinction*. New York: Columbia University Press.

———. 2018. "Extinction." In *Critical Terms for Animal Studies,* edited by Lori Gruen, 169–81. Chicago: University of Chicago Press.

———. 2019. *The Wake of Crows: Living and Dying in Shared Worlds*. New York: Columbia University Press.

van Styvendale, Nancy, J. D. McDougall, Robert Henry, and Robert Alexander Innes. 2021. "Introduction: 'Art for Life's Sake': Approaches to Indigenous Arts, Health, and Well-Being." In *The Arts of Indigenous Health and Well-Being,* edited by

Nancy Van Styvendale, J. D. McDougall, Robert Henry, and Robert Alexander Innes, 1–20. New York: Routledge.

Vaughn, Audrey K., M. Nils Peterson, William R. Casola, Kathryn T. Stevenson, and Lara B. Pacifici. 2021. "Using the Implicit Association Test to Evaluate Subconscious Attitudes Toward Snakes." *Anthrozoös* 35, no. 2: 293–306.

Vilhauer, Ruvanee P. 2009. "Perceived Benefits of Online Support Groups for Women with Metastatic Breast Cancer." *Women & Health* 49, no. 5: 381–404.

Vint, Sheryl. 2010. *Animal Alterity: Science Fiction and the Question of the Animal*. Liverpool: Liverpool University Press.

Vivian, Bradford. 2018. "Memory: Ars Memoriae, Collective Memory, and the Fortunes of Rhetoric." *Rhetoric Society Quarterly* 48, no. 3: 287–96.

Voluntary Human Extinction Movement. "About the Movement." *The Voluntary Human Extinction Movement*, accessed February 24, 2022. http://www.vhemt.org/aboutvhemt.htm

von Mossner, Alexa Weik. 2017. *Affective Ecologies: Empathy, Emotion, and Environmental Narrative*. Columbus: Ohio State University Press.

von Uexküll, Jakob. 2010. *A Foray into the Worlds of Animals and Humans with a Theory of Meaning*, translated by Joseph D. O'Neil. Minneapolis: University of Minnesota Press.

Vonnegut, Kurt. 2019. *Slaughterhouse-Five or The Children's Crusade: A Duty-Dance with Death*, 50th Anniversary Edition. New York: Modern Library.

Wallace-Wells, David. 2019. *The Uninhabitable Earth: Life after Warming*. New York: Tim Duggan Books.

Walsh, Catherine. 2010. "Development as Buen Vivir: Institutional Arrangements and (De)colonial Entanglements." *Development* 53, no. 1: 15–21.

Weber, Andreas. 2017. *Matter and Desire: An Erotic Ecology*. White River Junction, VT: Chelsea Green Publishing.

Weber, Andreas. "Why Erotic Ecology?" *The Biology of Wonder*. Accessed February 23, 2022. https://biologyofwonder.org/matter-desire-0/why-erotic-ecology

Wheeler, Wendy. 2014. "Natural Play, Natural Metaphor, and Natural Stories: Biosemiotic Realism." In *Material Ecocriticism*, edited by Serenella Iovino and Serpil Opperman, 67–80. Bloomington: Indiana University Press.

Whitson, Steve, and John Poulakos. 1993. "Nietzsche and the Aesthetics of Rhetoric." *Quarterly Journal of Speech* 79, no. 2: 131–145.

Wildlife Center of Virginia. 2022. "The Problem with Feeding Ducks." *The Wildlife Center of Virginia* accessed February 23, 2022. https://www.wildlifecenter.org/problem-feeding-ducks

Wiley, Stephen B., and Jessica Elam. 2018. "Synthetic Subjectivation: Technical Media and the Composition of Posthuman Subjects." *Subjectivity* 11, no. 3: 203–27.

Willingham, Emily. 2022. "How Facebook Is Saving Snakes." *Scientific American*, September 1, 2022. https://www.scientificamerican.com/article/how-facebook-is-saving-snakes/

Wills, Brandon K., Michael Billet, S. Rutherfoord Rose, Kirk L. Cumpston, Francis Counselman, Kendall J. Shaw, Kellen G. Cresswell, and Nathan Charlton. 2020.

"Prevalence of Hematologic Toxicity from Copperhead Envenomation: An Observational Study." *Clinical Toxicology* 58, no. 4: 262–65.
Wilson, Denise, Amohia Boulton, and Isaac Warbrick. 2019. "Physical Wellbeing of Māori." In *Routledge Handbook of Indigenous Wellbeing*, edited by Christopher Fleming and Matthew Manning, 71–85. New York: Routledge.
Winthrop-Young, Geoffrey. 2010. "Afterword: Bubbles and Webs: A Backdoor Stroll through the Readings of Uexküll." In *A Foray into the Worlds of Animals and Humans with a Theory of Meaning*, edited by Jakob von Uexküll, translated by Joseph D. O'Neil, 209–43. Minneapolis: University of Minnesota Press.
Wolfe, Cary. 2010. *What Is Posthumanism?* Minneapolis: University of Minnesota Press.
Woods, Heather, and Leslie Hahner. 2019. *Make America Meme Again: The Rhetoric of the Alt-Right*. New York: Peter Lang.
Yates, Amanda Monehu. 2021. "Transforming Geographies: Performing Indigenous-Māori Ontologies and Ethics of More-Than-Human Care in an Era of Ecological Emergency." *New Zealand Geographer* 77, no. 2: 101–13.
Zagacki, Kenneth S., and Victoria J. Gallagher. 2009. "Rhetoric and Materiality in the Museum Park at the North Carolina Museum of Art." *Quarterly Journal of Speech* 95 no. 2: 171–91.
———. 2018. "Rhetoric and Materiality in the Museum Park at the North Carolina Museum of Art." In *Readings in Rhetorical Fieldwork*, edited by Samantha Senda-Cook, Aaron Hess, Michael K. Middleton, and Danielle Endres, 81–98. New York: Routledge.
Zamir, Tzachi. 2012. "Without Offending Humans: A Critique of Animal Rights" [Review]. *Notre Dame Philosophical Reviews*. https://ndpr.nd.edu/reviews/without-offending-humans-a-critique-of-animal-rights/
Zuboff, Shoshana. 2019. *The Age of Surveillance Capitalism: The Fight for a Human Future at the New Frontier of Power*. New York: PublicAffairs.

Index

About the Author

Cynthia Porter Rosenfeld, MSW, MS/PhD, is a teaching assistant professor at North Carolina State University. Rosenfeld has contributed to books on posthumanism, digital humanities scholarships, and the ethics of using virtual reality in classroom settings, in addition to several scholarly articles that get at the question *How do we live well with others in a more-than-human world?*

www.ingramcontent.com/pod-product-compliance
Ingram Content Group UK Ltd.
Pitfield, Milton Keynes, MK11 3LW, UK
UKHW041837031026
15587UKWH00002B/829

9 781666 915464